The Ethnography of Tourism

ANTHROPOLOGY OF TOURISM: HERITAGE, MOBILITY, AND SOCIETY

Series Editor

Mission Statement

The Anthropology of Tourism: Heritage, Mobility, and Society series provides anthropologists and others in the social sciences and humanities with cutting-edge and engaging research on the culture(s) of tourism. This series embraces anthropology's holistic and comprehensive approach to scholarship, and is sensitive to the complex diversity of human expression. Books in this series particularly examine tourism's relationship with cultural heritage and mobility and its impact on society. Contributions are transdisciplinary in nature, and either look at a particular country, region, or population, or take a more global approach. Including monographs and edited collections, this series is a valuable resource to scholars and students alike who are interested in the various manifestations of tourism and its role as the world's largest and fastest-growing source of sociocultural and economic activity.

Advisory Board Members

Books in Series

The Ethnography of Tourism: Edward Bruner and Beyond, edited by Naomi M. Leite, Quetzil E. Castañeda, and Kathleen M. Adams

Capoeira, Mobility, and Tourism: Preserving an Afro-Brazilian Tradition in a Globalized World, by Sergio González Varela

Rethinking the Anthropology of Love and Tourism, by Sagar Singh

Tourism and Wellness: Travel for the Good of All?, edited by Bryan S. R. Grimwood, Heather Mair, Kellee Caton, and Meghan Muldoon

Bourbon Street, B-Drinking, and the Sexual Economy of Tourism, by Angela Demovic

Anthropology of Tourism in Central and Eastern Europe: Bridging Worlds, edited by Sabina Owsianowska and Magdalena Banaskiewicz

Apprenticeship Pilgrimage: Developing Expertise through Travel and Training, by Lauren M. Griffith and Jonathan S. Marion

Cosmopolitanism and Tourism: Rethinking Theory and Practice, edited by Robert Shepherd

Tourism and Language in Vieques: An Ethnography of the Post-Navy Period, by Luis Galanes Valldejuli

Tourism and Prosperity in Miao Land: Power and Inequality in Rural Ethnic China, by Xianghong Feng

The Ethnography of Tourism

Edward Bruner and Beyond

Edited by Naomi M. Leite,
Quetzil E. Castañeda, and
Kathleen M. Adams

LEXINGTON BOOKS
Lanham • Boulder • New York • London

Published by Lexington Books
An imprint of The Rowman & Littlefield Publishing Group, Inc.
4501 Forbes Boulevard, Suite 200, Lanham, Maryland 20706
www.rowman.com

6 Tinworth Street, London SE11 5AL, United Kingdom

British Library Cataloguing in Publication Information Available

Library of Congress Cataloging-in-Publication Data Available

Library of Congress Control Number: 2019949409

ISBN 9781498516334 (cloth)
ISBN 9781498516341 (electronic)

Contents

List of Figures

Acknowledgments

This project emerged in the context of the American Anthropological Association's Anthropology of Tourism Interest Group, which organized a series of panels for the 2014 AAA annual meeting to mark the fortieth anniversary of the first AAA session on tourism (Mexico City, 1974). One of those panels, first proposed by Quetzil Castañeda and organized by Naomi Leite with Quetzil as chair, focused on Ed Bruner's contributions to the anthropology of tourism. Although two of the original six panelists were unable to provide a chapter for this volume, the remaining four form the kernel of this book.

Quetzil and Naomi would like to thank our fellow ATIG board members for their support of the original panel and the book project, and for their dedication to innovative research. We are especially grateful to the original panelists, whose papers helped to spark this volume, and to the eight additional contributors who joined us to round out the book's topical coverage and stayed the course throughout the long journey to publication. Kathleen Adams graciously came on board as a third editor in a moment of crisis. Her expertise, energy, and steadfast support of this project have been invaluable.

In preparing this book we have been fortunate to be assisted by an entire cast of talented individuals. At SOAS, University of London, Naomi's graduate students Liselle Milazzo, Tayler Ulmer, and Nicholas Ogris painstakingly transcribed several hours of interviews with Ed Bruner recorded in 2014 and 2015, and Brooke Armstrong prepared the book's bibliography. In Illinois, Daniel Sanchez helped immensely with the cover design, sifting through seemingly endless photo possibilities, narrowing the selection, and preparing them for submission. Susanne Belovari, Archivist for Faculty Papers at the University of Illinois Urbana-Champaign, scanned a set of Ed Bruner's photos for inclusion in the volume. Meg Wallace provided expert copyediting, and Kaitlyn Rabach assisted with the index. We are especially grateful to series editor Michael Di Giovine, who was the first to embrace this project and remained enthusiastically engaged throughout its lengthy gestation, and to Kasey Beduhn, our acquisitions editor at Lexington Books, who has the patience and kindness of a saint. Her steadfast backing at crucial moments has been vital.

In addition, Naomi and Kathleen would like to extend heartfelt appreciation to Paula Mota Santos, the host of the 2018 Buroquinha "Writing the World" retreat, as well as to our fellow participants. Thanks to the wisdom, generosity, and good cheer of Paula and our fellow traveler Jade Lee, our week spent in Portugal crafting portions of this book was not only productive but filled with spectacular meals, midday swim breaks, and even our very own unicorn horns (not to mention a few memorable sunset gin and tonics at the end of long writing days).

Of course, each of us has incurred our own personal debts of gratitude along the way.

Kathleen Adams: My thanks to Naomi and Quetzil for bringing me into this project in early 2018. It has been an honor to work with them. The flurries of book-related communications launched two lovely new friendships. I also thank my dear, ever-wise, life-long friend Andrew Causey for his helpful feedback on an earlier draft of chapter 7. In addition, I thank Michael Di Giovine, Paula Mota Santos, Janice Nadler, and Denise Adams for advice and support at crucial moments. I am also grateful to my husband, Peter Sanchez, for his intellectual and emotional support, as well as his solo parenting while I was away at the Buroquinha writing retreat to work on the volume. (I know the souvenir bottle of aged Porto I brought him from Portugal is inadequate compensation for his unending patience and good cheer.) Finally, I extend my deep gratitude to Daniel Sanchez for his work on the book cover. Though the final product is not his (due to press constraints), his artistic eye, photograph selection, and generosity with his time were deeply appreciated.

Naomi Leite: My thanks to all of the contributors for their thoughtful scholarship, openness to editorial feedback, and commitment to seeing this volume through. It is difficult to express how grateful I am to Quetzil for the project's original inspiration, as well as his unstinting dedication to it as it morphed over the years from our original six-person panel to the fully fledged book now before us. At times the entire project seemed the baton in a ceaseless relay race, passed back and forth between us; I know how lucky I am to have had him there to carry it when I was unable. Kathleen was a marvel as she rapidly came up to speed and ran a length in our stead at a difficult time. I am blessed to have them both as co-editors and friends. Paula Mota Santos, Nelson Graburn, David Picard, Valerio Simoni, Michael Di Giovine, Jade Lee, Adam Clayton, and the Little Engines That Can writing group offered additional camaraderie, thoughtful critique, and moral support. This book is a monument to intellectual community of the very best kind—and to the friendship it fosters. Lastly, I thank my husband, Jason Head, for every single step along the way. He knows why, and how much.

Quetzil E. Castañeda: I am deeply grateful to everyone involved in the book as authors and interlocutors. *Felicidades* to all of you for your fortitude in bringing this project to completion. It seems like each one of us, authors and editors, endured at least one familial, personal, professional, or medical crisis—all swirling around and between us in various discordant cycles. I especially have appreciated collaborating with Naomi and Kathleen, who have always made the journey a particularly rewarding and enriching adventure. I look forward to the next journey with you both. I also want to express my deepest appreciation to Lisa Breglia, Juan Castillo Cocom, Edy Dzidz Yam, Allan Maca, Bogdan Popa, and Christine Vassallo-Oby, for your enduring friendship, engaged support, and *chen tsíikbal*—personal, intellectual, and always *extra*-vagant. Finally, I dedicate my contribution to this book as well as my project of studying Chichén to my mother, who passed away on September 10, 2018, in the last month of finalizing the manuscript. I don't remember much about my first trip to Chichén Itzá in 1972 or '73, but I do remember the now lost photo that my father took of my mother and I at the edge of the Sacred Cenote of the Magicians of Water. This image of two tourists posing as my mother and I is the origin of my work there and here.

Part I

ORIENTATION

Introduction

The Ethnography of Tourism

Encounter, Experience, Emergent Culture

Naomi M. Leite, Quetzil E. Castañeda, and Kathleen M. Adams

> Tourism has less to do with what other people are really like and more to do with how we imagine them to be, and in this respect is like any other form of representation, including ethnography.
>
> —E. Bruner 1989b, 439

This is a book about a particular approach to the study of tourism, one that coalesced in the 1990s and continues to be prominent today, but that has roots reaching deep into the history of anthropology: the experience-near, interpretive-humanistic perspective. This approach is often (though not exclusively) associated with the work of the American anthropologist Edward Bruner. As the epigraph for this chapter suggests, it is characterized by an attentiveness to representation, imagination, interpretation, meaning, and the inherent subjectivity of ethnography itself, elements we address in the pages that follow. More broadly, this is a book about the ethnographic study of tourism, including its historical rise, its current state, and its possible future directions. Using Bruner's work as a lens and a fulcrum, contributors to this volume draw upon their own field materials to shed light on a series of core methodological and theoretical concerns in contemporary tourism ethnography.

In this introductory chapter, our aim is not to offer an overview of the history of qualitative tourism studies or the anthropology of tourism, a task ably taken up by numerous other scholars.[1] Instead, we focus on *ethnography*, in the anthropological sense of theories and methods grounded in in-depth, face-to-face, typically long-term encounters between researchers and the people, places, and phenomena they study. The chapter is divided into three parts. The first introduces anthropological ethnography and its transformations over the past several decades, highlighting the significance of those changes for the ethnographic study of tourism in general and the approach to tourism ethnography associated with Bruner in particular. In the second part, we provide a detailed introduction to Bruner's interpretive-humanistic perspective. We highlight the origins of his research agenda, particularly the concerted efforts he and others made in preceding decades to bring lived experience, interpersonal encounter, and emergent culture to the forefront of the anthropological agenda. This section also provides a guide to key concepts and themes addressed in Bruner's work, including authenticity, narrative, contested sites, the touristic borderzone, and embodiment. Finally, in the last part we offer a view of the ethnography of tourism beyond Bruner, via a critical assessment of his enduring contributions to the field as well as an overview of the chapters that make up this volume.

ETHNOGRAPHY, ANTHROPOLOGY, AND TOURISM

The word *ethnography* refers both to a research method and to a kind of text, an attitude and a way of knowing. Although its meaning varies across the disciplines, for anthropologists—the primary proponents of the term over the past century, and the focus of this book—to *do* ethnography in the classical sense is to undertake field research into how people live and make sense of the world around them. This is achieved *participant observation*, anthropology's hallmark method of living, traveling, eating, shopping, working, and generally participating in the day-to-day life of the people we study, while simultaneously analyzing our experiences through the triple lenses of social theory, history (change over time), and comparison with analogous phenomena elsewhere. In pursuit of holistic understanding, anthropological ethnographers immerse themselves in the myriad cultural, social, and environmental elements affecting human experience and material existence in their field sites, taking seriously the words and actions of the people with whom they interact.

If all goes according to plan, the result is *an* ethnography, a text (often a monograph) that provides a detailed description and multifactorial analysis of some aspect of life for participants in that society. These texts are typically organized around a particular empirical topic or theoretical question: What does being an adult female entail for a given cultural group, and how is the transition from one life phase to another marked through ritual? What can auto plant workers' shop-floor conversations, gift exchanges, and afterwork pursuits teach us about changing models of success in a globalizing economy? How do international human rights discourses and regional gender norms combine to produce policies that directly shape life prospects for female refugees? What does it mean to be "modern"?

While the kinds of questions anthropologists ask have changed continuously over the years, the ethnographic mode of conceiving of research problems, studying them, and writing about them crystallized at a specific historical moment, the early decades of the twentieth century. When the foundational figures of social and cultural anthropology—Bronisław Malinowski, Franz Boas, Margaret Mead, and others—were developing their methods, there was considerably less movement of people and goods around the globe than there is today, and for the majority of the world's peoples lifelong stasis, rather than mobility, was the presumed norm. Although some early anthropologists undertook research in European settings (e.g, Arensberg 1937; Chapman 1935), that was very rare; the vast majority of ethnographic fieldwork in the first half of the twentieth century was carried out in lands far from the metropole, usually among indigenous populations or in colonial settings. Typically, the ethnographer would arrive after an arduous journey by train or ship, find local accommodations, and set about becoming integrated into the local community. There he or she would remain for a period of months or even years, and little sign of the external world could be expected to intrude—at least, not as represented in the resulting ethnography. When foreign travelers—traders, missionaries, military troops, speculators, colonial settlers, and, yes, eventually even tourists—appeared during their fieldwork, many anthropologists simply chose not to mention it in their writing. Their focus was on understanding the internal dynamics of the local community.[2]

Ethnography and Tourism: A Changing Relationship

We rarely have a record of when tourists first turned up in any given ethnographer's "'hidden corner' of the world" (V. Smith 1977, 49). We do know, however, when their presence made its way into the scope of ethnography. Beginning in the 1960s, a smattering of publications and calls for further research on tourism-related phenomena started to appear in anthropology journals. However, from the early 1960s to the early 1970s, the vast majority of these contributions came from anthropologists and sociologists who had stumbled upon tourism in the midst of researching something else (e.g., Nuñez 1963; Graburn 1967; 1969; Greenwood 1972; Cohen 1971; Ansari 1971; Aerni 1972; see also Leite and Graburn 2009, 39; Nash 2007). Those who subsequently made tourism a focus of their research often had to make up for years of willfully ignoring the tourists in their field sites. Pierre van den Berghe, a leading ethnographer of Latin American tourism and ethnic relations in the 1980s and 1990s (van den Berghe and Keyes 1984; van den Berghe 1994; see also Nash 2007), nicely captured the default practice of that earlier era while presenting his work at a 1982 seminar on ethnicity and tourism: he confided that in the 1970s he had become so adept at cropping the tourist hordes out of his field photos that he now lacked images to illustrate his lecture, which was an analysis of changing tourism dynamics at Machu Picchu, Peru.[3]

In this early period of ethnographic research involving tourism, anthropologists typically focused on "host" communities, meaning rural, indigenous, and postcolonial populations whose lands had become destinations for tourists. True to the classic model of ethnography, these anthropologists remained in place for an extended period, living side by side with their respondents through the ordinary activities of daily life. What had changed was that they now recognized tourists and tourism development as part of the mix. Early on, the questions these researchers asked followed lines of inquiry developed by anthropologists studying the effects of colonialism on indigenous culture: What was the observable impact of tourists and an emerging tourism economy on traditional lifeways? What signs of change, if any, could be identified in social structure, resource allocation, gender norms, and the annual cycle of labor, rest, and ritual? How were existing local dynamics of social stratification or exclusion affected by the presence of outsiders? More broadly, was international tourism development a solution for the woes of economically struggling, "underdeveloped" regions or nations, or did it lead to social, cultural, and environmental degradation?

Observers have described this early scholarship as understanding tourism "as an external force, something like a moving billiard ball, acting upon a pre-existing object, ethnic culture" (R. Wood 1998, 223). According to this model, tourism, like colonialism, would abruptly strike the local culture head on, scattering its traditional elements in all directions—that is, it would have an *impact*. Implicit here is an image of societies and cultures as fragile, unchanging wholes that must be protected from outside influence if they are to remain intact. Yet cultural contact, exchange, and influence are a constant of human existence; tourism is but one version of a perpetual process of change. Thus, as Quetzil Castañeda puts it, "there is absolutely no 'tourism impact' to study. It is the wrong concept by which to adequately understand the complexities of tourism" (Castañeda 1996, 9). It was not until the 1980s that more nuanced and flexible ways of thinking about the dynamic cultural phenomena surrounding tourism would fully flourish. Prior to the "opening up of anthropology" in the 1980s and 1990s—to use Edward Bruner's (1984a) phrase—most tourism research was framed explicitly in terms of impact, and freighted with concern over the loss of traditional lifeways around the globe.

Yet even the earliest ethnographic studies of tourism were more sophisticated and exploratory than the image of "impact" would suggest. The essays collected in *Hosts and Guests: The Anthropology of Tourism* (V. Smith 1977), the first major volume of ethnographic studies on the topic, are a case in point. While most are framed as an exploration of tourism's effects on postcolonial, indigenous, and rural societies, they simultaneously address complex questions still at the heart of contemporary anthropology. These include cultural commodification and the persistence or even revival of ethnic and cultural identification; tourism as a novel form of interaction in which strangers conduct transactions across substantial cultural difference, while exercising agency in choosing what to reveal and how much to engage; the place of power and creativity in the sale of souvenir arts and crafts; and the stereotyped images that toured populations develop of both their visitors and themselves. From the very beginning, then, anthropological ethnographies of tourism have highlighted locals' agency, recognized that societies were hardly stagnant prior to the tourists' arrival, and combined an interest in material aspects of social, cultural, and economic change with attention to questions of meaning and identity (Lett 1989, 276–77).[4]

Nonetheless, the dominant ethnographic and theoretical models of the era tended to pose challenges for early ethnographers of tourism, particularly those seeking to place the international traveler at the center of their lens. Given the continuing dominance of the classic model of immersive fieldwork, this should come as no surprise. How could one spend months or even years participating in the day-to-day lives of tourists, who are—by definition—a transient population? "Given these temporal conditions," explains Nelson Graburn (2002, 20), "the only way for the ethnographer to amass a suitably large amount of data would be to multiply repeat observations and questions with hundreds of tourists, but this strategy, of course, leads to qualitative data of no great depth." He continues, "ideally, no ethnographer would study persons taking part in a ritual or other short-term event and expect to be able to interpret them *meaningfully without putting them into the whole context of their continuing lives*, especially the meaning and mark that such events might (or might not) leave on their ordinary non-ritual lives" (2002, 20; italics in original).

In light of these limitations, many social scientists writing about international tourists in the 1970s and 1980s turned to theoretical and typological generalizations (e.g., Cohen 1972; 1974; 1979a; V. Smith 1977, 8–12). This approach, of course, precluded rich and nuanced ethnography. The sole chapter in *Hosts and Guests* that focuses entirely on tourists, for example, is decidedly non-ethnographic in character.[5] Written by Graburn (1977), it offers an overarching, cross-cultural theory of the meaning and role of leisure travel in the temporal flow of life, grounded in diverse examples drawn from an interdisciplinary literature.

Among those few who attempted ethnographic research on leisure travelers, most limited themselves to examining the behavior of domestic tourists—that is, those traveling to sites within their own country—as an extension of ongoing fieldwork on a related topic (e.g., Nuñez 1963; M. Smith 1982; Passariello 1983; Pfaffenberger 1983). Other researchers accompanied overseas travelers who shared their own national or cultural background, for example a Swedish anthropologist visiting the Gambia with a tour group from Scandinavia (Wagner 1977) or an American joining his compatriots on charter yachts in the Caribbean (Lett 1983).[6] It was only after a period of dramatic change to the discipline's methodological and theoretical underpinnings, running from the late 1970s through the mid-1990s, that anthropologists would find a way to undertake ethnographic research on diverse international tourists that could support their continued commitment to understanding "the native's point of view" (Geertz 1974)—whether that "native" be a resident of a rural tourism destination, an itinerant tour guide, or a tourist arriving from afar.

In the next section, we briefly sketch how the anthropological practice of ethnography and its use as a way of knowing have changed since the late 1970s. In particular, we emphasize the significance of those changes for the study of tourists and tourism. Especially important here is the rapid expansion of topics, materials, and methods considered to be legitimately within the discipline's purview. This transformation was sparked by the introduction of interpretive, postmodern, and poststructuralist theory into the anthropological mainstream. Bruner, today known largely for his influential work in the anthropology of tourism (e.g., E. Bruner 2005a), contributed to that earlier shift to a more humanistically oriented, experience-centered anthropology, well before he turned his attention to tourist performances, sites, and interactions. In fact, his turn to tourism research in the 1980s came relatively late in his nearly seventy-year career. Nonetheless, his approach to tourists and tourism—like that of his fellow travelers into tourism-related research, James Clifford (1988; 1997), Malcolm Crick (1985; 1989), Fred Errington and Deborah Gewertz (1989), Richard Handler (1988; see also Handler and Gable 1997), and Benetta Jules-Rosette (1984), among others—arose directly from his participation in the intellectual ferment and exploratory ethos of the preceding decades. His work thus provides an excellent example of a critical turning point in the ethnography of tourism.

"Opening Up" the Ethnography of Tourism: The Interpretive/Postmodern Turn

It is not an exaggeration to say that the ethnographic method and writing of the early twenty-first century would be unrecognizable to anthropologists of the early twentieth. Whereas earlier ethnographers focused on recording characteristic aspects of "the way of life of a people"—a tribe, an ethnic group, a rural village, an entire country—and depicted either a static, monolithic culture or a small-scale society undergoing transformation as a result of external forces, by the 1980s many had shifted their attention to questions of meaning, process, power, discourse, representation, social change, individual agency, and the nature of anthropological knowledge itself (E. Bruner 1990).

Various factors led to this transformation. First, a cadre of humanistically oriented scholars became interested in the work of symbolism and metaphor in society (e.g., Douglas 1966; Turner 1969; 1974; J. Fernandez 1986). Some focused their attention on the ways that events, performances, texts, and other expressive forms could be interpreted in their own right as meaning-making vehicles or as indications of primary preoccupations or values among the people studied (e.g., Geertz 1973; Bruner and Becker 1979; E. Bruner 1984b; 1986c; Babcock 1986; Turner and Bruner 1986). Second, a growing number of anthropologists began to reflect upon ethnography as a literary genre, a unique form of textual representation that could be analyzed in terms of common narrative structures, illustrative tropes, rhetorical modes of establishing authority, and interplay with other forms of cultural representation (Crapanzano 1977; Boon 1982; Fabian 1983; E. Bruner 1986b; Clifford and Marcus 1986; Clifford 1988; Geertz 1988; Van Maanen 1988). For some, this exploration provided an opportunity to break explicitly with the discipline's conventional norms, experimenting with such alternative genres as fiction, drama, and memoir for representing understandings culled from fieldwork (e.g. Wolf 1992; Visweswaran 1994; Behar and Gordon 1995). Many also began to experiment with incorporating analysis of their own interactions and emotional responses into their analytical tool kit and then using first-person narrative in the resulting text to make that aspect of ethnographic research explicit (e.g., Rabinow 1977; Myerhoff 1978; E. Bruner 1984a; 1993a; Rosaldo 1984; Kondo 1986).

This attention to the nature of ethnographic knowledge and voice emerged from the recognition that anthropologists are themselves culturally and socially situated beings, and hence always experiencing and interpreting phenomena from a particular subject position. Such reflexivity—reflection on one's own role in producing what had previously been accepted as ethnographic truth—led to a firestorm of debate over whether objectivity was possible, whether ethnographic research could ever be replicated and verified, and whether ethnographic texts could be divorced from the power relations, personal proclivities, and situated perspectives embedded in their production (Clifford and Marcus 1986; Kondo 1986; Marcus and Fischer 1986; Fox 1991; Behar and Gordon 1995).[7] As a discipline, anthropology became riven by two fundamentally different visions of the place for a researcher's emotional and interpretive consciousness in the ethnographic endeavor: either reflexivity, subjectivity, and individual experience were to be minimized because they would distort the facts and hence knowledge itself, or those very biases were methodological and analytical tools to be embraced as the social foundation of truth and an essential aspect of producing anthropological knowledge (Rosaldo 1989, 168–95).

The introduction of reflexivity in this methodological sense came somewhat late to the anthropology of tourism. Although the first wave of ethnographers clearly reflected on their own subjective positioning in their research, they did not use that reflexivity explicitly as a tool of analysis. In fact, while some of these scholars, including Valene Smith (2015; see also Nash 2007), drew on their subjective experiences as professional guides or travel agents, some recall doing so in an oblique fashion, lest their ongoing work in the tourism industry be perceived as compromising their legitimacy as scholars. Even so, the increasing acceptance of humanist and constructivist approaches in ethnography did not immediately lead tourism scholars to foreground their subjective experiences in their research and writing. In the ethnography of tourism, reflexivity in this form became fully accepted only via Bruner's contributions: as a long-established scholar publishing in high-visibility journals, he was well positioned to make broadly persuasive arguments for the value of reflexivity to tourism studies.

Scholarly concern with reflexivity arose amid much larger changes in the intellectual and political climate of the time. By the late 1960s and 1970s, the core certainties of the modern era—the inevitability of progress, the righteousness of colonialism and Western domination, the infallibility and moral clarity of science—were questioned and challenged, not only on American university campuses, but in broader public and international arenas. Student protests and campus demonstrations against the war in Vietnam, or against university investments in apartheid South Africa, in tandem with civil rights movements (feminism, the Black Panther movement, the American Indian movement, the Grey Panther movement and, from the early 1980s, the gay rights movement), challenged the singular authority of the modern period, ultimately giving way to the doubts, fragmentary narratives, and highlighting of positioned truths that came to characterize what was termed *postmodernity*. For our purposes, postmodernity is best understood as a widespread critique of approaches associated with modernity. The term is also used in reference to a historical era associated both with the rise of postcolonialism (a sociopolitical and intellectual critique of the continuing legacy of colonialism and maintenance of neocolonial institutions) and with capitalist transformations variously known as late capitalism, globalization, and neoliberalism.

For many anthropologists, these changes could not be ignored. On the one hand, scholars began to critically examine the role anthropology had played in processes such as nation building, colonialism, neocolonialism, and imperialism. Ethnographers could no longer view themselves as fundamentally separate from the people and phenomena they studied. On the other hand, technological changes in communications, transportation, mass media, and systems of representation heightened the condition of global interconnectivity, transforming the very experience of fieldwork. These changes also meant that in many parts of the world formerly silenced "natives" were finding new platforms that enabled them to speak for themselves to ever-broadening audiences. Ethnographers could no longer claim the unique authority to represent "other cultures."

The rise of speedier forms of transport and budget airlines also enabled rapid expansion of the tourism industry, bringing ever greater numbers of travelers from Europe, North America, and elsewhere to destinations that anthropologists had long claimed as their own domain. The stories, photographs, and objects these tourists brought back and shared in their home countries further challenged ethnographic authority and exceptionality, as did the rapidly multiplying media representations of far-off places, including educational documentaries, television travel shows, magazines like *National Geographic*, and memoirs, novels, and films about other cultural worlds.[8] Moreover, in response to the growing flood of curious travelers, "indigenous cultures around the world were reshaping themselves for presentation to a tourist audience" (E. Bruner 2012, 151). As more and more people, objects, ideas, and practices came unmoored from their points of origin and moved around the globe, the very concepts on which anthropology had been founded—above all, the concept of culture as a consistent, integrated whole, rooted in place, that defined and dictated the lives of others (Keesing 1990)—seemed to be breaking apart.

As Julia Harrison (this volume) shows, Bruner was a leading figure in the debates over reflexivity and postmodernism in the 1980s. These debates, in turn, paved the way for the transition he and others made into studying tourists and tourism. As one in a wave of anthropologists engaging with an interdisciplinary array of postmodern and poststructuralist thinkers such as Roland Barthes, Jean Baudrillard, Jacques Derrida, Umberto Eco, Michel Foucault, and Jean-François Lyotard, Bruner emphasized that "postmodern ethnography describes juxtapositions, pastiche, functional inconsistency, and recognizes, even celebrates, that cultural items originating from different places and historical eras may coexist. . . . Contemporary ethnographers no longer try to mask outside influences, nor do they see them as polluting pure culture" (E. Bruner 2001, 899). Instead, they turned their gaze back on themselves as one such "outside influence," exploring how their presence was interpreted by the populations they studied.

In the process, many found themselves reflecting on how ethnographers might resemble the tourists they had formerly dismissed (Crick 1985; Errington and Gewertz 1989; E. Bruner 1989b; 1989c). Indeed, in 1985 Crick went so far as to suggest that perhaps one reason tourism had not yet become a major focus for anthropologists was "precisely because tourists are relatives of a kind; they act like a cracked mirror in which we can see something of the social system which produces anthropologists as well as tourists. More than that, tourists remind us of some of the contexts, motives, experiential ambiguities, and rhetoric involved in being an anthropologist" (Crick 1985, 78). A range of anthropologists began to recognize that cultural tourism and ethnography are parallel, cross-pollinating systems of representation, each exploring, documenting, and consuming difference (e.g., Clifford 1988; E. Bruner 1989b; 1995; 2001; 2005a; 2012, 150–51; Castañeda 1996; Kaspin 1997).

Building on their humanist-oriented readings of postmodern and poststructuralist theory, Bruner and his fellow travelers sought to reorient both the focus of ethnographic research and the culture concept itself. Describing his efforts in collaboration with Victor Turner, Barbara Myerhoff, James Boon, Renato Rosaldo, James Fernandez, Barbara Babcock, and others (E. Bruner 1984b; Turner and Bruner 1986), Bruner wrote:

> We wanted to show how the newer processual postmodern theories actually worked in ethnographic practice. We thought of ourselves as not just talking about the postmodern project but actually doing it, utilizing the methods of ethnography. Moving beyond the questioning of ethnographic authority, which we accepted and appreciated, we stressed experience, performance, and narrative, as these concepts were based upon indigenous rather than the ethnographer's conceptualizations. (2005e, 206)

A few years later, in an essay reflecting upon his career, Bruner underscored what these changes had meant for the study of culture:

> The key point was to propose a revised concept of culture—to study culture not as it functioned, nor as a set of normative behaviors, but rather as it is *experienced*, to emphasize what is meaningful in people's lives. By "experience" we meant what emerges to consciousness, as opposed to what people say or do. . . . Culture was to be seen as being generated as it was expressed, as emergent and contested, in polyphonic interplay, as always in production. (2012, 149–50)

It should be noted that by the 1980s, arguments for this reconceptualization of culture were pervasive throughout the discipline and in the social sciences generally (see, e.g., Rabinow and Sullivan 1979). The "opening up" of anthropology occurred precisely because anthropologists of diverse intellectual, theoretical, and analytical visions—including new waves of feminist and Marxist scholars—were simultaneously reaching the same conclusions about the shortcomings of the classic concept of culture and the importance of reflexivity. Thus, Bruner's role is so important not because he singlehandedly originated a shift to a new understanding of culture, but because he was among the vanguard articulating these arguments in the 1980s and played a leading role in creating symposia and collaborative publications on the topic (see, e.g., E. Bruner 1984a; 1984b; 1986a; 1986c; 1989c; 1993a; Turner and Bruner 1986). When he subsequently shifted his focus to tourists and tourism, his eloquent emphasis on experience and emergent culture would render him one of the most cited anthropologists in tourism studies during the 1990s.[9]

We return to the concepts of experience and emergent culture, two hallmarks of Bruner's approach to tourism, in the next part of this chapter. First, however, there is more to be said about the methods of anthropological ethnography in a rapidly changing world. By the late 1980s, some ethnographers influenced by the changes of the previous two decades began developing methods of studying the lived experience of mobile populations, among them migrants, refugees, seasonal workers, and transnational families (e.g., Appadurai 1990; Ong 1993). They also examined the lives of individuals who themselves stay put but are active participants in communities "whose bounds are limited by a social, not a territorial, field" (Kenny and Kertzer 1983, 6; Rouse 1991). Others sought to trace the circulation of ideology, imagery, objects, media, and strategies of governance across cultural domains and geographical regions (e.g., Martin 1994; Naficy 1993; Steiner 1994; Tsing 1993).

Describing such efforts in a programmatic review of the literature, George Marcus (1995) dubbed this nascent form of ethnography *multisited.* Anthropologists, he wrote, were moving "out from the single sites and local situations of conventional ethnographic research designs to examine the circulation of cultural meanings, objects, and identities in diffuse time-space" (1995, 96). Multisited ethnography addresses translocal cultural systems and processes, rather than studying a limited set of individuals or a particular circumscribed geographical location. For example, in a study of the multiple meanings of West African masks as they move through different types of markets, a multisited ethnographer would follow the objects as they travel through venues of production, transit, display, representation, and sale, from the rural carver's workshop to the market stall in the nearest city to the regional dealer to a high-end gallery in New York. Some "sites" might be ephemeral or virtual, rather than physical—for example, discourses about "primitive art," histories of imagery, widely circulating stories about the mask trade, and the like. Not all locations involved would receive the same amount of time and attention from the ethnographer. Instead, because the focus of multisited ethnography is the condition of interconnectivity itself, understood through movement and traced as it is generated and experienced by the various actors involved, classical ethnography's mandate of long-term residential immersion does not apply (Marcus 1995, 100).

Not all anthropologists of tourism today would describe their methods as multisited ethnography. However, Marcus's statement on multisited fieldwork and its widespread, if tacit, endorsement by leading anthropologists paved the way for a younger generation of scholars to undertake ethnographic research on the move (Gupta and Ferguson 1997). Combined with the opening up of anthropology's scope, theoretical underpinnings, geographical range, and topics from the late 1970s to the 1990s, these methodological developments provided the conditions for a thriving new era of research on tourism, an era that would see it addressed as a system of representation, a locus of cultural production, and a multifaceted social field (Leite and Graburn 2009). Today, ethnographers interested in tourism focus their fieldwork not only on "host" populations or on the traveling unit, but on an open-ended array of spatially non-contiguous practices, images, sites, narratives, actors, and relations, including but not limited to those grounded in specific tourist destinations.

KEY CONCEPTS AND THEORIES: THE ETHNOGRAPHY OF TOURISM ACCORDING TO BRUNER

In some ways, Bruner is an unlikely focus for a book on the ethnography of tourism. While his tourism work is widely cited in anthropology and beyond, he published fewer than a dozen fieldwork-based articles on the topic, many of them reprinted in *Culture on Tour* (2005a). For the most part, his writing does not directly engage the theoretical or ethnographic contributions of other anthropologists studying tourism. Why, then, has his work been so influential? In this part of the chapter, we sketch the major theoretical and methodological elements of his approach and indicate some of the ways they have been taken up by other scholars. These include his overriding interest in experience and meaning; his emphasis on tourism encounters and emergent culture as an analytic focus; and his innovative vision of how to conduct ethnographic fieldwork on tourism. We then turn to his contributions to a series of concepts and themes that lie at the heart of tourism ethnography today, each of which is taken up and extended by one or more chapters in this volume.

A Focus on Experience and Meaning

> There are no raw encounters or naïve experiences since persons, including ethnographers, always enter society in the middle. At any given time there are prior texts and expressive conventions, and they are always in flux.
>
> —E. Bruner 1986c, 12

In a review of *Cannibal Tours*, Dennis O'Rourke's quirky 1987 documentary on primitivist tourism in Papua New Guinea, Bruner (1989b) argues that the film brings to the fore problems of practice, power, representation, and discourse that anthropologists and other tourism researchers had hitherto ignored. To underscore the point, he describes a telling moment from the film: "A New Guinea elder says, 'We don't understand why these foreigners take photographs of everything,' which is a very good anthropological question" (1989b, 441). He would return to that question—at base, how and what tourists "see," how local people interpret their looking, and what meanings all parties draw from the encounter—in many different guises over the ensuing decades, for it provided a point of entry into broader themes of anticipation and experience, cultural narratives and social identities, ambiguity and contestation that have become the hallmark of his ethnographic work on tourism.

Although Bruner's 1989 essay does not offer an answer to the New Guinea elder's question, we can find one possibility in a vignette he wrote two decades later, describing tourist behavior in a very different setting: "In Cairo on a crowded downtown street, with automobiles and busy pedestrians in modern Western dress going in every direction, there appeared a donkey cart driven by an old man wearing peasant clothes. A group of tourists rushed over to photograph the cart and its colorful driver, as it appeared quaint, possibly authentic, and somehow more 'Egyptian'" (E. Bruner 2011, 198). This mundane photographic encounter is significant, he writes, because it reveals the "selective perception" that shapes tourists' experiences: here, the cart "met the tourists' preconceptions of what they expected to find in Egypt, but it was the tourist expectations that constructed their experience by selective perception. The tourists themselves made their experience congruent with their expectations" (2011, 198). In other words, through the narrowed focus of a camera lens these tourists momentarily created the Egypt they had come to see, the destination they imagined long before the trip began. The rest of the scene receded from view.

Bruner was not alone in highlighting the contrast between the tourist's selective vision and that of the everyday resident. Sociologists Dean MacCannell (1976) and John Urry (1990) also examined aspects of how leisure travelers focus their attention on different things, and in markedly different ways, than they would in ordinary life. For MacCannell, this is because tourists are looking for an authenticity lacking in their day-to-day existence: they want to encounter something "real," whether it be a rare, precious object (a piece of moon rock, the crown jewels, Elvis's guitar); a hidden pocket of exotic culture or untouched nature; the site of important historical events; or a place to experience the simpler life of long ago. The tourist's pursuit is aided, MacCannell suggests, by a process he calls *site sacralization*: the transformation of ordinary objects and landscapes into attractions, simply by virtue of their being marked in some way (1989, 41–43, 109–33). The actual marker may be on-site, in the form of a sign, a brochure, or an explanatory placard, but sources of information produced and circulated elsewhere are equally important. In MacCannell's terms, guidebooks, travel blogs, newspaper articles, websites, maps, and even other people's photos and souvenirs are "off-site" markers that direct and focus the tourist's attention before, during, and after the tour. In the case of Bruner's elderly Egyptian and his donkey cart, we might surmise that everything from guidebooks and postcards to Hollywood films and *National Geographic*–style reportage contributed to what those tourists in Cairo hoped to see—and did see, focusing their cameras upon it despite the absence of any on-site marker at all.

For Bruner, as for MacCannell before him, tourist attractions are not just naturally there, awaiting their visitors; they are *made*, created through various forms of human intervention and selective perception. A tree struck by a bullet in the attempted assassination of a nineteenth-century world leader, MacCannell (1989, 128) tells us, is just a tree—until there is a marker to commemorate the event. Yet it also takes a certain attitude, a kind of consciousness, for a passerby to notice the marker in the first place. We have all had the experience of not noticing some element of our everyday surroundings until an out-of-town visitor points it out. For Urry (1990), this alternative consciousness is a defining feature of the *tourist gaze*, a term he coined to describe how leisure travelers experience objects and people differently than they would in non-touristic contexts:

> The tourist gaze is directed to features of landscape and townscape which separate them off from everyday experience. Such aspects are viewed because they are taken to be in some sense out of the ordinary. The viewing of such tourist sights often involves . . . a much greater sensitivity to visual elements of landscape or townscape than normally found in everyday life. (Urry 1990, 1)

In their 2011 reworking of the concept, Urry and coauthor Jonas Larsen (2011, 195–200) stress that the tourist gaze is not solely a matter of sight. It is a broader mode of interacting with people and place: it involves a heightened tuning of one's senses to the experience of *being there*. Which sense or stimulus takes the fore—sight, smell, touch, taste, temperature, sound—will depend on the tourist's expectations and the kind of destination involved, whether a tropical paradise, romantic Paris, or a site of ancestral trauma. It will also depend on the traveler's prior knowledge and unconscious assumptions, based on widely circulating imagery and ideas about other people and places.

Although Brunner, MacCannell, and Urry share an interest in tourist motivation and aspects of tourist experience, Bruner's approach diverges sharply from MacCannell's and Urry's in a crucial respect. The latter two formulated general theories, as did many other social scientists writing about tourism early on (e.g., Cohen 1979a; Graburn 1977; Gottlieb 1982). Bruner was among just a handful of scholars in the 1980s and early 1990s who eschewed grand theorizing to focus on ethnographic study of how individual tourists (and other stakeholders) experienced specific sites, performances, and encounters, as well as the diverse meanings they drew from the visit.

For example, in his (1996a) study of tourism at Ghana's Elmina Castle and its dungeons—built by the Portuguese in 1482, used for centuries as a staging area for slave ships, colonized first by the Dutch and then by the British, and now a UNESCO World Heritage Monument—Bruner conducted extensive participant observation among various groups engaged with the site. He also conducted interviews, convened focus groups, and examined a wide range of representations of this tourist destination, including fiction and memoir. His multifaceted approach enabled him to tease apart the site's diverse meanings for Dutch, British, African-American, and Ghanaian tourists, professionals working in tourism and heritage preservation, and local residents of disparate backgrounds. In the resulting article, he argues that each group has an interest in the castle, but for different reasons; and they are concerned with different periods in the castle's long history, in effect rendering it several different destinations in one. Such a multiplicity of meanings, Bruner suggests, is equally possible at any tourist site, because of variations in what different visitors anticipate, imagine, and encounter while on tour (see also E. Bruner 1993b; 1994; 2005b; 2011). Put another way, not every tourist in Cairo would pause to photograph that picturesque peasant driving his donkey cart; indeed, some might not even notice him, their expectations having predisposed them to focus on something else altogether. It is only through careful ethnographic fieldwork among diverse groups of tourists that one can identify which is which.

Bruner's long-standing interest in the nature of human experience, and how anthropologists can study it, is another factor differentiating him from his contemporaries in tourism research. In 1979–1980, years before he began studying tourism, Bruner and Victor Turner organized a symposium for the American Anthropological Association on what they dubbed the *anthropology of experience*. In his introduction to the resulting edited volume (Turner and Bruner 1986), Bruner emphasizes that the book's contributors, like other anthropologists in the interpretive-humanistic tradition, are not attempting to document culture in the abstract sense of "generalized routines, clusters of customs, norms, habits, and prevalent patterns" (E. Bruner 1986c, 4). Instead, their focus would be on specific incidents and events, moments involving expressive forms like narratives, dramas, or the visual arts, as a means to explore how individuals *experience*—that is, subjectively engage with, interpret, and draw meaning from—the world around them.[10]

Bruner and his colleagues drew from philosophers like Wilhelm Dilthey, who wrote that "reality only exists for us in the facts of consciousness given by inner experience" (Dilthey 1976, 161). Pointing to the dynamic relationship between "reality (what is really out there, whatever it may be), experience (how that reality presents itself to consciousness), and expressions (how individual experience is framed and articulated)," Bruner argued that the differences between these three components of human experience should be a primary target of ethnographic research (1986c, 6). The tripartite model of reality/experience/expressions would subsequently reappear in his writing on tourism: "the trip as lived, as experienced, and as told are never exact replicas of one another, and there is no precise mimetic correspondence" (2005b, 19). Yet they are inextricably linked, for "experience structures expressions and expressions structure experience" (1986c, 7; see also 2005b, 19; 2011, 199–200). Furthermore, our own "lived experience . . . is the primary reality" to which we have access (1986c, 5). This insight has important methodological implications. Because we cannot directly experience another's experiences, we must make inferences based on their expressions: conversation, story, gesture, imagery, display of emotion, and the like. Discerning the meaning of such expressions is always an act of interpretation, whether as an ordinary individual simply making sense of another person's words and behavior, or as a professional interpretive anthropologist examining cultural meaning.

Bruner's analysis of the multiple meanings of Elmina Castle offers a clear example of how one might use the reality/experience/expressions triad in ethnographic research. In addition to participant observation, Bruner's fieldwork included careful attention to relevant public debates, memoirs, newspaper articles, passing comments from visitors and locals, and the narrative arc of dramatic reenactments for tourists in the slave dungeons. These he interpreted as reflections of, and influences upon, diverse visitors' experiences. Together with his own sensory and emotional experiences of engaging with the material realities of the site, these many expressions of others' experience provided the evidence for Bruner to generalize about commonalities and differences in the meanings drawn by different groups. He also noted the embeddedness of those meanings in larger histories of slavery, ethnic relations, and social inequality in Ghana, the United States, and beyond. As he later reflected, "a site is not fully described from the actors' perspective, but must be seen in its larger political context as mediating between the global and the local" (E. Bruner 2005b, 12).

Other ethnographers undertook similarly immersive, reflexive, and experience-focused participant observation among tourists during the 1990s—for example, on package tours to Holocaust sites (Feldman 1995; Kugelmass 1992), roots pilgrimages to West Africa (Ebron 1999), and independent visits to such famed destinations as the Taj Mahal (Edensor 1998) and the pilgrimage route to Santiago de Compostela (Frey 1998). Like Bruner, these scholars examined diverse expressions that both reflected and contributed to tourist experience—among them memoirs, letters, stories heard and retold, photographs, visitor books, conversations on-site—and zeroed in on questions of imagination, embodiment, selective perception, and meaning. Nonetheless, Bruner's work has been by far the most widely cited. There are a number of possible reasons for this. One is that these scholars were often more junior, some of them doctoral students (e.g., Beer 1993; Louie 1996; Tucker 1999); at such an early career stage, their research and writing were less likely to draw attention. Bruner was already a well-known anthropologist who had been elected to several prominent national leadership roles.[11] Moreover, his studies of tourism arose directly from his earlier, widely read theoretical contributions on experience and expressive culture. As such, his contributions to tourism ethnography, typically published in leading anthropological journals, carried an in-built authority that sanctioned and smoothed the way for his contemporaries undertaking similarly humanistic and interpretive work on tourism.

Tourism as Emergent Culture

To many anthropologists of the 1960s and 1970s, international tourism represented a threat to traditional ways of life. Viewing culture as a unified whole that was vulnerable to contamination or degradation—via colonialism, tourism, or industrialization—they expressed grave concerns over shifts from local systems of exchange to participation in an international economy and the transformation of local traditions into performances for tourists. In the late 1970s and 1980s an alternative vision of culture began to take shape, wherein "the way of life of a people" was no longer thought to be fixed and timeless, but instead to be constantly in flux. From this perspective, culture could not be lost or protected, destroyed or preserved. It was always *emergent.*

The implications of this modified understanding of culture for the ethnography of tourism were enormous. If, as Bruner put it in the 1980s, "people construct culture as they go along and as they respond to life's contingencies" (1989c, 113), then tourism could be approached as just one more social space in which new meanings and experiences were generated. Attractions, performances, and souvenirs need not be dismissed as tacky imitations of "the real thing," but could be examined as meaningful incidences of cultural production. Moreover, from this perspective each tourist performance, visit to an attraction, or fleeting interaction would constitute a unique social context, prompting a unique experience for the visitor and for individual residents. In this light, "the final meaning for the tourists, locals, and producers is not given a priori but emerges in dialogic interplay during their interactions" (E. Bruner 2005b, 17).

As the above vignette of the New Guinea elder suggests, Bruner's model of the dynamic relationship between reality, experience, and expressions situates tourism both as a domain of activity in the world-out-there and as a system of representation. This system of representation involves a powerful feedback loop: tourists, like ethnographers, interpret what they see and hear, and yet their interpretations of the people and places they encounter are shaped by layer upon layer of prior representations of others' interpretations. At the same time, the diverse groups of actors engaged with tourist destinations—guides, performers, hotel workers, artisans, people living locally—bring their own assumptions, stories, memories, and expectations to bear in making sense of their encounters with tourists; and tourists' expressions of their experiences can have a profound effect on the self-image and subsequent self-presentation of toured populations (E. Bruner 1991; 1996b; 2001; see also Abram, Waldren, and MacLeod 1997). In the tourism encounter, myriad influences conjoin to produce, or *construct,* place and people, self and other.

Broadly speaking, this vision of tourism as a creative process—a process in which participants' experience of reality emerges as a cultural construction, in part through the influence of widely circulating narratives and imagery—is the cornerstone of what has become known as Bruner's constructivist, processual approach. As he states, "Performances for tourists arise, of course, from within the local cultural matrix, but all performances are 'new,' in that the context, the audience, and the times are continually changing. To put it another way, performance is constitutive" (2005b, 5). To say that performance is constitutive or that reality is socially constructed is not to suggest that there are no material realities, no larger political and economic forces at work, but rather to acknowledge the complex ways that they converge and blend with participants' culturally inflected imaginings, assumptions, expectations, perceptions, and senses to produce lived experience (Schütz 1962; Berger and Luckmann 1967; Garfinkel 1967; E. Bruner 1973; Graburn and Leite, this volume).

As Sally Ness (this volume) discusses in her chapter, constructivism focuses on the social and cultural production of meaning, experience, and self as continuously shaped and reshaped through an ever-changing array of factors. Here again, the relationship between reality as lived, as experienced, and as expressed is key. Do African American tourists *really* relive the long-dead slaves' experience in the dungeons of Elmina Castle (E. Bruner 1996a)? Are the dances and crafts on offer at Mayers Ranch, a Kenyan tourist destination, real examples of Maasai culture (Bruner and Kirshenblatt-Gimblett 1994)? Is the dominant historical narrative presented at Abraham Lincoln's New Salem, another site whose diverse meanings Bruner (1993b; 1994) has analyzed, true? Constructivists shift the terms of discussion, asking instead, by what criteria are they judged to be "real" or "true" in this particular context? From whose perspective? Through what series of interpretations? With what effects? The ethnographic task is thus one of learning whether the feelings and performances are real or the narratives true for specific individuals, as lived experience; how that experience takes form and evolves as people move through the destination and engage with one another; and how both the experience and the meanings attributed to it are culturally shaped.[12]

The (Touristic) Encounter

As a constructivist, Bruner has argued that tradition and heritage, like experience and meaning, are "not given before the touristic encounter but, rather, the encounter generates [their] definition" (Jules-Rosette and Bruner 1994, 405). But what, precisely, is a touristic encounter? The concept has long been central to ethnographic research on tourism, though it is rarely explicitly defined or even consistently used in the same ways with the same meaning (see, e.g., Karch and Dann 1981; V. Adams 1996; Conran 2006; Babb 2010; Simoni 2016b).[13] Generally speaking, scholars use the term at an individual, micro level to refer to moments of contact "between people, between people and space, amongst people as socialised and embodied subjects, [and] between expectations and experience" (Crouch 1999, 1). At a collective level, *the touristic encounter* is often used by analogy with *the colonial encounter* to refer broadly to a characteristic form of cultural confrontation between "hosts" ("natives") and "guests" (foreign travelers).

Most commonly, though, *touristic encounter* refers generically to any moment in which one or more travelers comes face to face with members of the local population or has a fleeting but meaningful engagement with some material or human aspect of the destination. Any such meeting involves political, cultural, moral, and economic dimensions that extend well beyond the individuals involved: "When the tourist-native encounter is referred to, one must not think of a person-to-person, one-to-one relationship, as if the focus were on an individual tourist interacting with an individual [native], for this masks the world system and discursive dimensions that frame the relationship, a frame set by Western discourse and by the hegemonic tourist narratives" (E. Bruner 1991, 242). Any encounter, that is, gains its meaning in no small part from macro-level power relations, narratives, discourses, and representations, in addition to each participant's prior imaginings and expectations.

In approaching the tourist encounter simultaneously as a unique moment of interaction between specific individuals and as a microcosm of larger systems of power and meaning, Bruner indirectly invokes the theoretical concept of encounter originally developed in symbolic interactionism, a strongly constructivist subfield of sociology. We know from Bruner's own accounts that he studied symbolic interactionism as a doctoral student at the University of Chicago and was drawn to its emphasis on the situational and interactive nature of social meaning (see, e.g., E. Bruner 1989c, 113; 2005b, 17; Di Giovine, this volume). Symbolic interactionists, most famously Herbert Blumer (1969), with whom Bruner studied, and Erving Goffman (1959; 1961; 1974), seek to understand how human beings make sense of situations and interactions as they arise. They approach this question with four essential premises in mind. First, individuals act toward things or people on the basis of the meanings those things or people hold for them; second, every interaction is shaped by a unique, unpredictable confluence of factors; third, the meaning of an interaction or event is not predetermined, but arises from on-the-spot, contextual interpretations that participants make of each other's behaviors and responses; and, finally, the meaning negotiated in any given interaction is always emergent, or coming into being, and thus liable to change (Denzin 1992; Fine 1993). In symbolic interactionism, *encounter* is a technical term for the basic unit of social organization within which an interaction takes place: it refers to "any joint act between two or more persons where the interactants are maintaining symbolic, visual, auditory, and at times tactile contact" (Denzin 1974, 270). In studying an encounter, the research goal is to tease apart how meaning is made, communicated, and understood, by "exploring its relationship to the larger social structure in which it is embedded" (E. Bruner 1963, 1416). A church service, a game of tag, or indeed a guided tour could each be analyzed in this light.

In a review of *Encounters*, Goffman's (1961) short book on the subject, Bruner (1963) argued that anthropologists would do well to follow the symbolic interactionists' lead. To do so, he wrote, would require systematic attention to micro-level encounters as meaningful events in their own right—a theme he and sociologist Jules-Rosette (1994) would underscore some thirty years later explicitly in relation to tourism. Urging social scientists to adopt a "more processual, fluid, and reflexive" mode of analysis, they argued that "the touristic encounter is less monolithic, more dialogical, more layered, more complex, more divergent, more processual, more experimental than previous scholarship has indicated . . . [It] must be considered as a moving border zone, a zone of creativity and emergence" (Jules-Rosette and Bruner 1994, 404, 406; see also E. Bruner 2005b, 18). This is the symbolic interactionist model of encounter, with all its indeterminacy, unpredictability, and emergent meaning. Thus, although Bruner mentions symbolic interactionism only rarely, its influence on his approach is profound. Whether in the form of tour group visits to packaged cultural performances or direct interactions between individual tourists and local people, encounters constitute Bruner's primary analytical focus, and capturing their emergent meaning, his goal.[14]

Over the past several decades, an interdisciplinary array of tourism ethnographers have used *encounter* in ways very similar to Bruner, as an analytical context for examining the production and negotiation of meaning (e.g., Tucker 1997; 2011; Crouch 1999; Crouch, Aronsson, and Wahlström 2001; Crouch and Desforges 2003; Leite 2017; D. Picard 2011; Simoni 2016b). Contributors to this volume, too, place the meaning-making force of encounters at the heart of their analyses, from informal interactions between tourists and locals in Cuba (Simoni) and negotiations over purchases between textile weavers and tourists in Guatemala (Little), to the different ways returning Toraja migrants and their descendants experience the sensorial rush of visiting their rural Indonesian homeland (Adams) and even an anthropologist-turned-tourist's challenging confrontation with the rugged natural landscape of California's Yosemite National Park (Ness).

Fieldwork: Site and Method

In the 1980s and 1990s, Bruner's use of encounters as the focal point of analysis offered a new model for how ethnographers might solve the seemingly intractable problem of long-term, immersive fieldwork among tourists. If one embraces the tenets of symbolic interactionism, as he did, then any given encounter can be approached as a microcosm in which diverse histories, discourses, representational structures, and systems of power and meaning combine with factors unique to the situation to produce new meaning. If this is so, then the encounter itself provides a coherent ethnographic field site. The ethnographer's immersion comes not from length of time among a particular group of people, but instead from multifaceted, ongoing, in-depth engagement with incidents of a given type of encounter. To clarify the point, here we review Bruner's own description of his methods and note how his approach diverges from that of others working in a similarly humanistic-interpretive vein.

In May 2005, Bruner published a brief essay in *Anthropology News*, the newsletter of the American Anthropological Association, as part of a series on the theme "Where do you do your fieldwork?" He answers the prompt with a deceptively simple statement: "In my tourism research I position myself in two different fieldwork locations, one with the tourists and the other in the travel destination" (2005f, 15). He goes on to argue that the ethnographer's task is to understand the meaning and nature of the tourists' journey, as experienced and as told; to that end, he suggests, one should accompany the traveling unit throughout the entire trip, beginning to end. He notes that he accomplished this by serving on multiple occasions as a tour guide or an academic lecturer, in addition to visiting the destination as an ordinary tourist, because "tourists were more likely to ask questions, to offer impressions, and to express their feelings directly to the tour guide than to another tourist" (2005a, 257n1; see also E. Bruner 1995; 1996b; 2005b; and Harrison, this volume). His research in the travel destination, on the other hand, focused on "the objects of the tourist gaze" (2005f, 16), via "traditional fieldwork . . . by staying in one place, sometimes for many months, to study the local performers and producers" (2005a, 257n4). There, his aim was to grasp "all the meanings of the site and to understand the perspectives of all relevant actors"—not only the tourists, but "the performers, the locals, tour agents, travel guides, business persons, officials in the national tourist bureaus, as well as those representing the airlines, hotels, and other international stakeholders in the tourist industry" (2005f, 16). He concludes that by conducting fieldwork both on the move with tourists and tourism professionals (guides, promoters, and so forth), and in an extended period of stationary participant observation at the tourist destination, he was able to develop a nuanced understanding of the multifaceted ways diverse stakeholders imagined, interacted with, and understood the destination.

Generally speaking, this two-pronged model of fieldwork accurately captures Bruner's methods. However, it glosses over what was most innovative and, ultimately, most influential in his approach to tourism ethnography. Bruner's research goal was neither a detailed ethnography of the traveling unit, nor a community study of the local population. His goal was an exploration of emergent culture in the encounter between the two. Thus, his focal unit of analysis was fundamentally different from that of most tourism ethnographers, both before and since. This is true even of those anthropologists who have similarly inhabited the position of tour guide as a central component of their research (e.g., Bunten 2008; 2015; Feldman 2016). For whereas the latter have been interested in capturing the lived experience of the (native) tour guide, and especially the practices involved in engaging tourist audiences and representing people and places, Bruner used his guiding role as a point of entry for in-depth participant observation in tourism *encounters*, as experienced from the points of view of tourists and tourism professionals. Moreover, his stationary fieldwork in the travel destination was hardly "traditional," as he claimed, even in the expanded sense afforded by the disciplinary transformations of the 1980s and 1990s. He did not live among the local population for an extended period, participating, for example, as a "host"—as did Nancy Frey (1998) in her stint working in a hostel on Spain's Camino de Santiago pilgrimage route, or Adam Kaul (2004), who pulled pints for tourists in an Irish village pub. Nor did he take on the role of a local performer, as Margaret Sarkissian (2000) did as an accordionist in a folkloric band playing Portuguese-Malaysian music for tourists in Malacca's "Portuguese Settlement."

Instead, whether traveling with tourists or remaining in place, Bruner's participant observation centered entirely on what transpired in the space of social interaction he called the *touristic borderzone*: the multiple forms of representation that converged to shape the encounter, the imaginings tourists and local people brought to it, the content and nature of their interactions, and the ways they spoke about it afterward, not to mention his own embodied experiences and interactions on site. His unit of analysis was thus neither the tour as a whole, nor the work of the guide, nor the life of the toured. It was the *tourist production.* As he explained in a 2015 interview with Michael Di Giovine, "I studied tourist productions. By this I mean a dance drama, a craft village, a reconstructed village or community. . . . My way of doing it was to take [the] tourist production as a performance and as a ritual and to study it in all its dimensions and its multiple layers of meaning. I tried to analyze the production from the beginning to the end, and how it moved through time and space" (see Di Giovine, this volume; also E. Bruner 2012, 151–52).

Bruner's emphasis on close analysis of tourist productions—presentations, folkloric dances, heritage sites, exhibits—emerged directly from the earlier efforts he and his colleagues had made to craft an anthropology sensitive to experience, meaning, and emergent culture, and it fit squarely within a broader shift in tourism ethnography toward interpretive, postmodern, and humanistic approaches. His approach was also greatly influenced by his interdisciplinary collaboration with performance studies scholar Barbara Kirshenblatt-Gimblett (Bruner and Kirshenblatt-Gimblett 1987; 1989; 1994; Kirshenblatt-Gimblett 1998). The two co-led an undergraduate study abroad program during the 1984–1985 academic year, taking their roving classroom to Japan, Bali, India, Kenya, Egypt, and Israel and visiting numerous tourist attractions and performances along the way. Although neither had previously studied tourism, together the two observed each tourist production as emergent culture, something to be taken seriously, a meaningful event that could be studied "much as any contemporary ethnographer would study ritual, even though these productions were not indigenous ceremonials but were designed for a foreign audience" (E. Bruner 2012, 151; 2005b).

In keeping with the analytical tools of performance studies, and particularly its overarching metaphor of cultural production as theater, Bruner argued that meaning lies not in the tourist production itself, but in people's interpretations of it (1993b, 23; 2005e, 203). Hence the only way to locate that meaning is to focus on individual encounters. For Bruner,

> it is only in the specifics of a particular performance that we may learn how the site and [its] story are actually produced and experienced. . . . Without a specified audience narratives have no meaning, because the meaning is only in the audience's reading of the text. This privileging of the specific leads to a consideration of the complexity of forces and the multiplicity of voices and meanings at work. (1993b, 14; see also 1984a; 2005c)

It is perhaps for this reason that Bruner's work has proved especially congenial for tourism studies scholars grounded in other disciplinary traditions.

Figure I.1. Edward and Elaine Bruner interviewing Maasai performers at Mayers Ranch with Jane Mayers translating, 1984.
Photo credit: Edward M. Bruner.

His focus on fine-grained analysis of specific tourist productions might seem superficially similar to that of scholars who "do ethnography" by visiting a site once or by viewing a performance and chatting with or distributing questionnaires to others in attendance. However, to mistake Bruner and like-minded anthropological ethnographers as doing the latter type of research is to overlook the depth and breadth of participant observation involved, the investment of time required, and the careful compilation of multiple lines of evidence to capture diverse perspectives: traveling alongside multiple tour groups and joining in their experiences, listening to individual tourists' stories during and after the journey, staying in place and repeatedly attending performances, spending time with performers before and after the show, observing and then hearing their accounts of what goes on, and simply living in the tourist destination. Above all, this approach to tourism ethnography entails a steadfast commitment to that time-honored anthropological goal of grasping the natives' points of view, in all their variety and even mutual contradiction.[15]

The ethnographic material presented in this volume moves far beyond the bounds of the formal tourist productions that Bruner studied, whether it be Canadians' stories about life at their summer cottages (Harrison); instances of strategic hospitality directed toward diverse types of foreigners in Madagascar (Picard); the embodied experience of hiking the arduous Yosemite Falls trail (Ness); or the interwoven events, stories, and representations of Mexico's Chichén Itzá at play in regional and national politics (Castañeda). Methodologically, however, contributors to this book share with Bruner a reflexive, experience-near approach to tourism ethnography, one grounded in careful, detailed fieldwork and attuned to inconsistencies, ambiguities, and resonances across diverse expressions of human meaning.

Throughout his writing on tourism—much of which is republished in his book of collected essays, *Culture on Tour* (2005a)—Bruner draws upon the theoretical grounding and ethnographic methods outlined above to reveal processes through which experience and meaning emerge and diverge among different populations at diverse sites and cultural performances: in Indonesia, Kenya, Israel, China, and the United States, on package tours and among independent travelers, with both domestic and international tourists. In addition to his widely influential emphasis on experience and emergent culture in moments of encounter, several of the primary concepts and preoccupations of contemporary tourism ethnography—authenticity, narrative, contested sites, the touristic borderzone, embodiment—either arose from or were substantially developed in his work. In the remainder of this section, we address each of these themes and concepts in turn. Two further concepts, identity and mobility, are covered in chapter 2 and so are not discussed here.

Authenticity

Authenticity has been a prominent theme in tourism studies for more than half a century (Leite and Graburn 2009, 43–44; Wang 1999; Cohen 2007). This is in large part because it is a perennial topic of discussion among tourists, guides, and heritage professionals, but also because it is the focus of long-standing theoretical debates in the field. Although the term has been defined in numerous ways and the theoretical and empirical questions raised vary widely, scholarly debate over authenticity can be grouped into three broad areas: tourist motivation and the tourism industry's response; the effects of cultural commodification; and the inherently contextual and constructed nature of "authenticity" itself.

Early on, social scientists sparred over whether tourism was best analyzed as the pursuit of shallow entertainment or as a serious search for meaning. For Daniel Boorstin (1962), tourism offered the prototypical example of the modern era's mindless consumption of *pseudo-events*—inauthentic, contrived experiences and attractions. For MacCannell (1973; 1976), it represented the opposite, a serious quest for something more authentic—pristine, untouched, "real"—than was available to Western tourists alienated from their rootless, mechanized middle-class life. MacCannell argues that the tourism industry is aware of the tourists' desire and responds with *staged authenticity*, simulating original cultural products, performances, and interactions as if they are not performed for financial gain. The tourists, ever suspicious that what they are being shown is fake, are doomed to continue their search.

Many ethnographers have responded to these generalizations by showing that there is significant variation across types of tourists and destinations. Not all tourist attractions are contrived, not all tourists are alienated, and not all tourists seek authenticity (Cohen 1979a; 1988; Wang 1999). Indeed, some travelers, dubbed "post-tourists," are said to take great delight in fakery, for example playfully embracing the tackiness and humor of the Las Vegas Strip with its overblown imitations of Venice and Paris (Feifer 1985; Urry 1990). Others simply hold different criteria for enjoyment and are perfectly satisfied with cultural performances that are not "traditional" but nevertheless offer "a good show" (E. Bruner 2001, 883; 2005b, 3). Still others seek a form of authenticity that lies in neither the destination nor the people visited, but in the social, emotional, and embodied experiences they hope to have along the way (Selwyn 1996, 21–25; Wang 1999; Harrison 2003; Conran 2006; Cohen 2007).

Authenticity also features in debates over the impact of cultural commodification (Cohen 1988). A common argument has it that when local lifeways are made into tourist attractions they lose their intrinsic meaning, becoming pale, inauthentic copies performed by exploited natives (e.g., Greenwood

1977). Yet ethnographic work has revealed that demand for packaged performances and visitable heritage may in fact serve to perpetuate traditional cultural knowledge, landscapes, and practices that would otherwise be lost (McKean 1977; Crick 1989; Azarya 2004). "Performing oneself" for tourists can also provide leverage for cultural recognition and political representation (Greenwood 1982; Hiwasaki 2000; Sylvain 2005). Particularly for participants in indigenous-owned and operated enterprises, the presence of an appreciative audience can foster pride in and identification with marginalized or even long-abandoned practices, while also generating economic and political benefits; in such cases, indigenous tourism workers and their tourist visitors may value the "authenticity" of what is on display equally (Bunten 2008; Bunten and Graburn 2018; Giraudo 2018).

Both of these debates stem from what Ning Wang (1999) calls an objectivist understanding of authenticity: a given cultural element or object is real or fake, preserved or lost. For at least one side of the debate or the other, or for the tourists they describe, "culture" is a shared cluster of practices, crafts, knowledge, folklore, and language whose elements are more or less authentic, more or less true to "tradition," and open to verification as such by elders or scholars with sufficient expertise. In this view, there is a "genuine" culture somewhere, hidden or in danger of being lost entirely, and its self-conscious enactment for an audience suggests it has already been lost.

The intellectual upheavals of the 1980s and 1990s introduced an altogether different angle on the question of authenticity, one that Bruner wholeheartedly embraced and developed in his work. If, as social scientists in the constructivist tradition argue, culture is not static but always emergent in the enactment, then no single practice or performance can be more or less authentic than any other (Greenwood 1982; Cohen 1988; E. Bruner 1989c; 1996b; 2001; 2005b; Crick 1989; V. Adams 1996; K. Adams 1997). Hence authenticity itself is a social construction, a quality variably attributed or contested by specific actors, in response to specific situations, using specific criteria. As Bruner puts it,

> For ethnographers, tourists, and indigenous peoples the question to ask is not if authenticity is inherent in an object, as if it were a thing out there to be uncovered or unearthed, but rather the question becomes, how authenticity is constructed. What is the process by which an ethnography, a tourist performance, or an item of culture or practice achieves an aura of being authentic? (E. Bruner 1989c, 113)

The immediate context is especially significant here. Noting that debates over the authenticity of objects, practices, or people arise only when someone or something has called their genuineness into question and claimed authority to authenticate, Bruner suggests that "ethnographers may then ask what segment of society has raised a doubt, what is no longer taken for granted, what are the societal struggles, and what are the cultural issues at work" (1994, 408; see also M. Picard 2008; Cohen and Cohen 2012). This grounded approach informs his study of divergent ways that tourists, museum professionals, historians, staff, and volunteer interpreters engage with the past at the New Salem Historic Site, a reconstructed nineteenth-century village and outdoor museum that aims to present an "authentic reproduction" of life at the time of Abraham Lincoln (E. Bruner 1994). While all of these stakeholders may be concerned with historical "authenticity," they understand that term—and consequently experience the site—in notably different ways.

By turning his attention to the empirical, ethnographic questions of when and why evaluations or claims of authenticity are made, by whom, according to what criteria, and with what broader repercussions, Bruner sidesteps the objectivist view of authenticity altogether, declaring it "a red herring" (2005b, 5). Instead, he and likeminded scholars propose that we view authenticity as "a struggle, a social process" (1994, 408; see also Gable and Handler 1996; Selwyn 1996, 18–28; Macdonald 1997; Kirshenblatt-Gimblett 1998). This constructivist approach is taken up and extended in the present volume by Walter Little and Valerio Simoni, both of whom explore instances in which emergent understandings of authenticity are constituted via interactions between tourists and residents. In the book's concluding chapter, Mary Mostafanezhad and Margaret Swain address diverse ways the debate has been furthered in recent decades by scholars in interdisciplinary tourism studies, particularly with regard to questions of representation, voice, and time.

Narrative

Bruner is strongly associated with the use of narrative as an object of ethnographic study, a methodological tool, and a form of ethnographic representation. In a 1986 book chapter titled "Ethnography as Narrative," he noted that all ethnographies are "guided by an implicit narrative structure, by a story we tell about the peoples we study" (E. Bruner 1986b, 139). Drawing on examples from Native American studies, he emphasized that "narratives change, all stories are partial, all meanings incomplete; . . . with each new telling the context varies, the audience differs, the story is modified" (1986b, 153). At the time, he had already been experimenting—as had several of his contemporaries—with a more explicit form of ethnographic storytelling: narrating a single event from fieldwork and using it as a focal point for analysis (e.g., E. Bruner 1973; Geertz 1973; Myerhoff 1978). His long-standing interest in stories and cultural narratives, particularly their partial, shifting, and often contested nature, informed and shaped his subsequent methodological focus on the role of narrative in the dynamic interplay of meaning and experience at tourist sites and in specific encounters (E. Bruner 1991; 2005b; 2005c; Harrison, this volume). Here, we briefly summarize three distinct ways that narratives—which he variously refers to as stories, tales, narrative, metanarratives, and master narratives—figure in his tourism ethnography, then address the methodological place of narrative in his work.

First, Bruner stresses that tourists anticipate, imagine, experience, and reflect on their travels through narrative—that is, through stories in the broadest sense. He distinguishes three types on the basis of their temporal position along "the touristic cycle": *pretour narratives*, the preconceptions travelers hold about the destination and the experiences they envision having, based on stories they read and hear before embarking; *on-tour narratives*, the stories tourists hear, read, and tell while en route, whether in interactions with guides, guidebooks, local people, or fellow tourists; and *post-tour narratives*, the anecdotes and travelogues they share after returning home, including those recounted only to themselves in memory (E. Bruner 2005c; see also 2005b, 19–27). These three types are not temporally separate in terms of shaping the tourist's experience: the pretour narrative inevitably colors what one notices and experiences during the tour, just as the experience of traveling may be shaped by anticipation of how the trip will be described to others in post-tour tellings.

Second, throughout his work Bruner has been attentive to how official narratives contribute to the meanings tourists and other stakeholders derive from the site, as well as how they become internalized as a result of the encounter. One especially instructive example is his analysis, written with Phyllis Gorfain, of the story of the mountaintop fortress of Masada in southern Israel (Bruner and Gorfain 1984). A famed tourist destination, the site is firmly associated with the AD 70 Jewish revolt against Roman rule, which resulted in the destruction of Jerusalem. According to the Masada legend, hundreds of Jewish fighters escaped to the fortress, where they managed to hold back Roman incursions for three years before committing mass suicide to avoid surrender. Bruner and Gorfain examine how this story has been told and retold in a variety of contexts, showing that its character and emphasis shift depending upon the teller's perspective on historical and contemporary politics in the region: it can be a heroic story of Jewish resistance in a hostile land, or a cautionary tale of the folly of absolutist, isolationist choices. For the tourist, physical presence on the mountaintop is a crucial component of coming to identify with the meaning he or she draws from the legend—whether that be the authoritative, heroic version or the counter-reading offered by the political left. As Bruner and Gorfain explain, "We move through the story as we move through the site. . . . We return not just with *the* story of Masada but rather with *my* experience of Masada. . . . A historical narrative is transformed into a personal narrative as we gain the right to tell the story. The consequence is to align individual biography with tradition, to incorporate national stories within the self" (Bruner and Gorfain 1984, 72–73; see also E. Bruner 2005b, 24).

A third way that narratives figure in Bruner's ethnography is in the overarching narrative structures, or cultural "scripts," that give meaning to travelers' otherwise disparate, potentially confusing experiences (Leite 2014, 266; Picard, this volume). Bruner illustrates this narrative structuring with a moment from his fieldwork as a guide. One rainy day, he took a small American tour group down a river in Thailand in an open boat. For protection they were accompanied by a Thai guard, who sat in the back holding a rifle. As the rain poured down, "the tourists were visibly nervous and somewhat disoriented until one said, 'This is like the movies about the Vietnam War along the Mekong [River].' The statement became the frame for the rest of the journey, for with that guiding narrative the tourists were transformed into American soldiers scanning the river bank for Vietcong" (2005b, 20). The tourist's invocation of a familiar storyline enabled participants to change their response to the ominous guard and the miserable weather, transporting them into a landscape of intrigue and playfully imagined danger.

Not all structuring narratives are so explicitly invoked. Bruner also highlights the generalized metanarratives about travel that shape one's vision of the journey, offering generic storylines like embarking on a voyage of discovery, time travel, a pilgrimage, or even the outdated vision of an anthropological quest for the vanishing primitive. Other narratives are more tethered to specific destinations. These draw on a combination of metanarratives and what Bruner terms master narratives (Hawaii as tropical paradise; Peru as land of the ancient Incas), the kind that circulate in tourism marketing, travel brochures, films, popular culture, and other media, motivating tourists and shaping their visions of what they will find (E. Bruner 1991; 2005b; see also K. Adams 1984; Selwyn 1996; Skinner and Theodossopoulos 2011; Salazar and Graburn 2014). As a result, a tourist's experience of the destination will have as much to do with overarching preoccupations and desires in his or her own society as with the site itself. Nonetheless, as a constructivist Bruner emphasizes that preexisting narratives do not determine the significance each individual will find in a destination (see Ness, this volume): meanings are always shaped by a unique confluence of factors, including prior life experiences, expectations, mood, weather, the people present, group dynamics, current events, chance interactions, and so on.

Although these terms—*narrative, master narrative, metanarrative*—also appear in other theoretical traditions, such as poststructuralist theories of text and discourse, Bruner's approach differs in three fundamental ways. First, he shows how narratives operate in specific cultural settings, drawing on the tools of ethnography, rather than invoking or analyzing them in the abstract as fixed texts. Second, and following from the first, Bruner's approach to narrative prioritizes the act of narration—that is, of storytelling, whether verbally or implicitly—in particular contexts, thickly described. Third, his primary interest in narratives is how they influence individual consciousness and, therefore, lived experience. In other words, for Bruner the formation, transmission, and contingent meanings of narratives are all objects of ethnographic description and analysis.[16]

Narrative also plays a role in Bruner's research methodology. In his fieldwork, he gathered stories told to him by multiple stakeholders in the tourist production—before, during, and after the encounter—in order to capture their diverse points of view (see Harrison and Ness, both in this volume). Believing that experiences and their expressions continuously shape each other, he argued that ethnographers should view narrative not just as a means by which people make sense of their own lives, but as essential fodder for understanding that sense-making anthropologically:

> One can never know directly what another individual is experiencing, although we all interpret clues and make inferences about the experiences of others . . . As anthropologists, our first responsibility is to respect people's accounts of their experiences as they choose to present them. We may not necessarily accept their claims or representations, but we had better understand them. (E. Bruner 1984a, 7–9)

It is no doubt for this reason that Bruner also makes liberal use of stories in his ethnographic writing, not only as data for analysis, but as a means of conveying and illustrating his arguments. And, like other humanistic-interpretive ethnographers, he also recounts stories of other people's storytelling as a means of deepening his readers' understanding.

Every chapter in this volume grapples with narrative in one way or another. Four are especially noteworthy for their extended engagement with Bruner's writing on the topic. In his chapter, David Picard explores ways that local people draw upon master narratives of self, other, power, and (spirit) possession to make sense of their interactions with diverse groups of foreigners—not only tourists, but also Catholic priests, doctors, and conservation scientists. Addressing a very different kind of narrative, Castañeda's contribution examines how layer upon layer of stories told in and about a world-famous tourist site point outward, speaking not only to a tourist audience but to participants in political and ideological conflicts extending well beyond the site. Harrison and Ness, too, explicitly engage Bruner's work on narrative, zeroing in on his use of stories in ethnographic research, analysis, and writing. Harrison narrates her personal intellectual trajectory in dialogue with Bruner's, addressing the ways in which stories have been at the center of both his and her own research and writing; for Ness, the narrative of her own experience of a single day's hike provides both the ethnographic material and the structuring trope of her argument, bringing the reader with her through the very meaning-making processes she describes.

Contested Sites

Narrative, experience, and emergent meaning also take center stage in Bruner's approach to tourist destinations as *contested sites.* For Bruner, there will always be multiple, often conflicting interpretations of a site, based on divergent values and interests both within and across different groups of stakeholders (E. Bruner 2005b). His analyses of Masada, Elmina Castle, Lincoln's New Salem, and Kenya's Mayers Ranch, for example, show how local and national officials, scholars, staff, and various populations of tourists and residents bring different narratives to bear in making sense of the site and determining the stakes of their engagement with it. Whether the narratives involved are specific to the location—as in competing representations of life at New Salem or opposing political interpretations of the Masada legend—or more general, as in the master narratives about slavery and postcolonial history in play at Elmina Castle, these points of contestation are never solely about the site itself. As Bruner puts it, "it is precisely in the struggle over meaning in historic sites, museums, and tourist attractions that the diverse segments of a democratic society are able to express their interests and stake their claims" (1993b, 23–24). Moreover, as he points out, the process through which one interpretation achieves authoritative status over others has much to do with political and economic power dynamics.

As Castañeda (this volume) observes, Bruner's emphasis on tourist sites as inherently emergent and multiple, in the sense of being interpreted and experienced differently by a wide range of actors, was a major contribution to the ethnographic study of tourism. Understanding that conflicts in meaning are natural, intrinsic, and inevitable enabled him to focus on the diverse factors generating different meanings for different audiences, without needing to assert that there was or was not culture loss involved or that one interpretation of the site was more valid or accurate than any other. Equally important was his recognition that seemingly site-specific conflicts—from debates over the content of explanatory texts and who should have the right to tell the story, to disagreements about appropriate activities in a given heritage site and what kind of historical restoration, if any, should be done—are often skirmishes in far larger political and social struggles.

Bruner's distinction between soft and hard contestation is particularly useful in this regard. He designates Lincoln's New Salem as an example of soft contestation, a conflict "not in the sense of a grand political conflict between colonial powers and the colonized, or between state power and the revolutionaries," but instead "a soft struggle over meaning between the 'official' interpretation of the site and how that official view is undermined by the processes of its own production" (1993b, 14–15). Masada, on the other hand, he deems an example of hard contestation, because the meanings at stake have direct geopolitical consequences: "People fought and died over control of the physical Masada, and they are fighting and dying today for what Masada means. . . . The past merges with the present and time frames become blurred as the question is posed, then and now: Should the Jews fight or accommodate?" (2005b, 11–12; Bruner and Gorfain 1984). While the precise terms in which it is invoked have changed since Bruner studied the site—Masada has shifted, perhaps, to a softer form of contestation—more recent ethnographers report that for professional guides it continues to present complicated, politically charged terrain (Kelner 2010, 62–64; Feldman 2016, 145–46).

Bruner's concept of contested sites has broad applicability. Although he applied it only to analyses of neatly bounded historical destinations, all of them clearly demarcated as tourist attractions, with professional staff, explanatory placards, and at least some degree of authoritative narrative framing, other ethnographers have taken a similar approach in a heterogeneous array of sites. Some, for example, have examined how amateur walking-tour guides of contrasting political persuasions construct entirely distinct experiences of a city through their choice of routes, the stories they tell, and the types of tourists they attract (Dahles 1996; Clarke 2000; Santos 2017). Others look more closely at the dynamic coproduction of the destination through interactions between guides or local people and tourists, emphasizing encounters where competing and suppressed narratives are in play (Tucker 1997; Chronis 2005; 2012; Leite 2007; Santos 2012).

Contributors to this volume enable us to extend the concept of contested sites further still. Kathleen Adams's chapter, for example, follows different generations of Toraja migrants on return visits to their homeland, exploring the physical and emotional "disconnects" they experience in relation to their evolving sense of identity and belonging. Here the visit itself is a contested site, with struggles over meaning arising between visitors of different generations, between migrant-tourists and their local kin, and between the travelers' anticipated and actual experiences of the encounter. Castañeda's chapter takes the idea of contested sites in yet another direction, looking beyond struggles over the meanings of Chichén Itzá to examine diverse ways the site has been mobilized in a range of conflicts involving and extending beyond its geographical borders. Finally, in their concluding chapter Mostafanezhad and Swain expand the discussion to include approaches from human geography and interdisciplinary tourism studies.

The Touristic Borderzone

Perhaps Bruner's most influential contribution to the ethnography of tourism is his concept of the *touristic borderzone*. Building on his earlier work on intercultural encounters, he developed the term in reference to the culturally constituted space in which local people—as performers, craft sellers, and the like—come into contact with tourists, not haphazardly, but "in structured ways in predetermined localities for defined periods of time" (E. Bruner 2005b, 17; 1996b; also Adams, chapter 2, this volume). Less a physical location than a context of encounter, the borderzone is best understood, in Bruner's words, as "a point of conjuncture, a behavioral field [conceived] in spatial terms" (2005b, 17).

The concept has three essential characteristics. First, the parties who interact in this behavioral field engage with it in very different ways. For tourists the borderzone is a realm of leisure, but for locals it is a place of work. Moreover, while each group moves in and out of the borderzone—"the tourists who come forth from their hotels, and the local performers, the 'natives,' who leave their homes to engage the tourists" (2005b, 17)—they are not equally mobile. The resident population stays anchored in place, witnessing "a recurring wave of travelers" passing through, whereas individual tourists are present only for a brief period, typically never to return (1996b, 158). Second, as a field of interaction centering around cultural representation, the borderzone is "a creative space, a site for the invention of culture on a massive scale," and therefore worthy of careful study (1996b, 159). Third, as with all touristic encounters, even the most scripted performances in the borderzone generate manifold meanings for diverse participants. Thus, in Bruner's work the borderzone is the temporal and behavioral space within which the tourist production takes place. Marked off from everyday life for both the tourists and the local people, it constitutes a distinct unit of activity for all involved, at least in the types of contexts Bruner studied—Maasai performances in Kenya, Balinese and Javanese dance dramas in Indonesia, and African-American tourist visits to Elmina Castle, among others. By definition fleeting, the borderzone in this view is the primary setting in which tourists, local residents, and the destination itself come together in the coproduction of meaning.

Subsequent ethnographers have extended the concept of the borderzone to address a broader array of contexts. For some, Bruner's claim that "natives have to break out of their normal routines to meet the tourists" in the borderzone (1996b, 158) misses the extent to which tourism has become enmeshed in everyday life for residents of destinations worldwide. As Little puts it in his (2004) study of Maya handicraft vendors in Guatemala, "in [the city of] Antigua the normal routine of local people *is* tourism," and the entire city may be seen as a borderzone of touristic engagement (2004, 87–88). Simoni (this volume) takes the concept in a different direction, examining the complex interplay between Cubans' long-term aspirations and their pursuit of informal encounters with tourists. Here the borderzone is indeed a distinct behavioral realm, but an unstructured one, rife with ambiguity, that contains within it the seeds to its transcendence: these locals and their foreign visitors sometimes form romantic relationships that both parties hope will exceed the limits of the borderzone and persist into a different life together. Picard (this volume) also explores how local residents engage with the borderzone for purposes that go well beyond the tourist encounter, in this case by using hospitality to appropriate foreigners' presumed power for their own self-empowerment in other domains. Further discussion of the borderzone—its origins, applications, and similarities to approaches developed by ethnographers in other disciplinary traditions—can be found in the chapters by Graburn and Leite, by Adams, and by Mostafanezhad and Swain.

Embodiment

In the 1990s ethnographers began to grapple with the role of the body in tourist experience (e.g., Edensor 1998; Frey 1998). Previously, sight had been taken to be the most significant sense involved—as the term *sightseeing* suggests—and with it an implicitly male gaze (Leite and Graburn 2009, 42). Overviews of the anthropology of tourism often point to Soile Veijola and Eeva Jokinen's "The Body in Tourism" (1994) as a turning point, as it was the first publication to systematically demonstrate the absence of the body from major works in the field and to argue for an experiential, multisensory approach to touristic encounter. While the essay certainly paved the way for a new trajectory of scholarship on the gendered tourist body and the importance of senses other than sight, it neglected a more subtle form of embodiment already at the heart of Bruner's work: the central role of the tourist's physical presence at the destination, or what he describes as "the sheer materiality of being there" (2005b, 24), in moving beyond the pretour narrative and into lived experience. As we saw in his analysis of Masada, for Bruner it is by moving *through* a site, experiencing it with all the senses, that the tourist comes to incorporate its meanings into his or her individual biography.

This concept of embodied encounter must not be mistaken as a simply physical phenomenon of seeing the site in person, as opposed to in a photograph or on a website. At Masada, for example, many tourists choose to make the exhausting two-hour trek up the winding path to the mountain's summit before touring the fortress site and imagining the siege. Here, "the physical exertion maximizes the experience" (Bruner and Gorfain 1984, 73). Similarly, writing of tourist encounters with the slave dungeons at Elmina Castle, Bruner details the smells and sounds, the touch of surfaces, even the bodily positions visitors are placed in as they move through spaces where earlier generations suffered so terribly (1996a; 2005b, 25). In this way, he shows how tourism's fleeting, ever-shifting conjuncture of bodies, narratives, imagination, and emotion serves "to inscribe the pretour narrative within the body of the tourist" (2005b, 24; see also Crouch and Desforges 2003). It is precisely this sensory intensity that makes journeys to long-imagined, narrative-saturated sites so potent for tourists seeking connection to a particular historical event, a generalized past, or their own roots (Bruner and Gorfain 1984; E. Bruner 1993b; 1996a; 2005b; 2011; Leite 2005; Adams, chapter 7, this volume).

In their contributions to this book, Adams (chapter 7) and Ness each transport the concept of embodiment to a new environment and extend its analytical reach. Adams examines the "somatic dimension" of return visits to the Toraja homeland, where migrants and their families find that visceral responses to the tastes, sights, smells, and physical conditions of the visit reveal generational and cultural rifts within the traveling group. For Ness, at stake is her own bodily movement up an extremely challenging hiking trail. She describes an alternative, skeleto-muscular form of knowledge generated by the trek, made possible through the connected copresence of herself and Yosemite Valley. The experience provides a constructivist case study in which meaning emerges not from the convergence of narrative and physical presence, but instead in her prelinguistic bodily and emotional encounter with the material demands of the landscape. This is embodiment at its most elemental, and it throws into sharp relief the validity of Bruner's assertion that tourist sites are themselves active agents in the construction of meaning (2005b, 25–26).

Our aim in this exploration of key themes and concepts in the ethnography of tourism, viewed through the lens of Bruner's work, has been to shed light both on the intellectual history of the field and on Bruner's role in its evolution. While it can be difficult to trace the origins of ideas and their subsequent trajectories of influence, the theoretical and methodological approaches to tourism outlined here—as encounter, as experience, and as emergent culture—clearly underpin a wide swath of scholarship in tourism ethnography today. Moreover, the continuing currency of both constructivism (particularly in regard to authenticity) and the concepts of narrative, contested sites, the borderzone, and embodiment suggests that the theoretical agenda laid down by interpretive-humanistic anthropologists like Bruner in the 1970s, 1980s, and 1990s continues to inspire scholarship in the present. Importantly, the concepts and approaches discussed here have evolved through the contributions of a great many thinkers, not all of them Bruner's collaborators, and as the afterword to this volume shows, they are continuously challenged, debated, and advanced across diverse disciplines. There can be no last word here, any more than there can be a single origin to enduring ideas.

BRUNER . . . AND BEYOND

No scholar or school of thought is beyond intellectual critique. While inarguably groundbreaking, Bruner's approach to tourism does have its limitations. In this regard, each contributing author to this volume delineates, responds to, and builds upon specific aspects of Bruner's work. In so doing, they provide critical commentary on his conceptual and methodological frameworks and highlight dimensions left underdeveloped. In this final section, we briefly review some of the critiques that have been leveled at Bruner's contributions overall. We then turn to an overview of this volume, outlining the material addressed and, where appropriate, queries and challenges raised by each chapter.

Critical Assessments

A common critique of Bruner's approach is that it affords only limited attention to political and economic forces that have shaped and structured tourist destinations. From the perspective of twenty-first-century anthropology and critical tourism studies, this is a shortcoming; there is no particular reason why a humanistic interest in meaning and experience should preclude systematic analysis of power dynamics, politics, and structural inequalities related to race, class, sexuality, and gender. Although Bruner's ethnographic work provides ample insights into experiential and discursive conflicts over representation, authority, meaning, identity, and belonging, and although he was certainly attuned to asymmetrical power relations involving race, class, gender, and so forth, more often than not these dynamics serve as contextual framings for his primary interests: lived experience, subjective meaning, embodied engagement, and the creativity of human expression in tourism's borderzones. To make an analogy, we might say that had he written about dance, he would have focused primarily on just one performance or a series of performances, emphasizing the dancer's embodied experience and the feelings and meanings generated among the audience, and making reference to social histories and broader systems of representation that inform them. He would not have offered a critical history of the dance form or an analysis of the political economy of dance as a profession or an exploration of racial or gendered barriers to participation.

Long before he turned to tourism, Bruner argued that ethnographers interested in experience and meaning should ground their research in specific instances of expressive culture—narratives, dances, dramas, rituals—because these were themselves a form of indigenous commentary: not only are they "people's articulations, formulations, and representations of their own experience," but they "leave the definition of the unit of investigation up to the people, rather than imposing categories derived from our own ever-shifting theoretical frames" (1986c, 9). It is in such expressions, he suggested, that a population's presuppositions and core values will be most apparent. As he later explained in relation to tourism, "A key advantage of studying a specific attraction in a single place is that the researcher can examine local and world politics ethnographically *as they manifest themselves in the site*, as opposed to resorting to facile overgeneralizations" (2005b, 12–13, emphasis added). The extent to which his work engages such local and world politics is debatable, a point he has made himself (2005b, 13); indeed, a primary focus on expressive forms can leave the ethnographer prone to downplaying or even overlooking the significance of structural questions of power and inequality, particularly if they are not explicitly or implicitly highlighted by participants.

In short, Bruner's primary ethnographic focus in studying tourism has been the "tourist production"—that is, the performance itself, both literally and in the anthropological sense of the socially situated enactment that constitutes all culture and meaning (Turner 1986; E. Bruner 2005b). Moreover, in much (though not all) of his work, the data arose from his travels as an anthropologist embedded within an organized tour group or as an independent traveler at tourist destinations. Thus, while we are given rich portraits of tourists' and guides' perspectives, we learn less about the ongoing quotidian, political-economic, and structural challenges facing people living in and around tourism destinations. In addition, the various meanings of these tourist productions are conveyed largely through brief yet memorable descriptions of places, interactions, and events, with only limited discussion of specific people. We lack the fleshed-out, ethnographically "thick" characterizations of participants and their lives that would convey more thoroughly the situatedness of their understandings and experiences. Hence, even as Bruner's body of work provides a persuasive argument for a deeply humanistic, interpretive, and experiential ethnography of tourism encounters, it does not offer a full-fledged example of what his approach could accomplish via long-term ethnographic fieldwork. Perhaps this is not surprising, given that his turn to tourism research came relatively late in his career, at a point when most anthropologists are no longer embarking on new ethnographic projects; *Culture on Tour* is, after all, a collection of mostly previously published, stand-alone articles based on brief research trips.

An additional result of Bruner's short-term fieldwork and narrow analytical aperture is that they do not permit him to trace concrete outcomes of the interactions and performances he observes, whether in individual lives or collectively for the various groups involved. Processual analysis, the exploration of cultural phenomena as they transpire moment by moment, can be extraordinarily powerful for understanding precise interactions and performances, but it does not easily translate to capturing long-term change. What practical consequences follow from the meanings tourists derive from a visit to Lincoln's New Salem (E. Bruner 1993b; 1994)? What conclusions do Maasai young men who perform for tourists as warriors (E. Bruner 2001) draw about their own life possibilities, and how does that relate to future choices?

Bruner hints at the possibility of lasting change, but does not address the processes through which embodied experiences and subjective meanings might give rise to something new. Given the fleeting nature of tourism interactions, to answer this question the ethnographer must move beyond the immediate setting—the borderzone of encounter—to follow the arc of individual lives as affected by broader political and social forces in which tourism is enmeshed (see, e.g., Brennan 2004; K. Adams 2006; D. Picard 2011; Lehrer 2013; Simoni 2016b; Leite 2017; Meiu 2017). This is as true of studying tourists and guides as it is for research on local people. While Bruner does advocate exploring how tourists integrate their experiences into their lives at home—for example, by attending tour group reunions and listening to tourist recollections prompted by their photos (2005b, 26–27)—these are more programmatic suggestions than they are thoroughly incorporated elements of his ethnographic research, and they have been more effectively pursued by others working in a similarly humanistic vein (e.g., Frey 1998; Harrison 2003; 2013; Louie 2004; Kelner 2010).

Finally, scholars have questioned Bruner's pithy generalizations about tourism, many of which are interpreted in a more sweeping sense than he evidently intended. His oft-quoted statements that "the tourist self is changed very little by the tour" (1991, 242), that tourists "really do not want an ethnographic perspective" and are not concerned with authenticity but will happily accept "a good show" (2005b, 2–6), and that "the natives have to break out of their normal routines to meet the tourists" in the borderzone (1996b, 158) have all been tested by ethnographers, including contributors to this book, in a range of tourism environments. Not surprisingly, his general findings about authenticity, identity, and the borderzone have been critiqued for their grounding in the narrow forms of tourism encounter he studied: upscale package tours bringing wealthy American and European travelers into contact with "native" performers, and domestic tourist visits to heritage sites. This is, however, a limitation he has openly acknowledged (2005b, 258n8).

It may be useful here to recall a disciplinary mantra common in the anthropology of the 1980s and 1990s: all ethnographic knowledge is partial. It is partial in the sense of being subject to the politics and dynamics of the era in which it was conducted, as well as to the particular interests and identity of the ethnographer. It is also partial in terms of being incomplete, regardless of how comprehensive an ethnographer aims to be. This partiality and perspectival specificity are both a strength and a weakness. Thus we can identify limits to Bruner's approach when it comes to rigorous analysis of the workings of power, politics, and history, as well as his method of situating himself primarily among tourists and, in most cases, only briefly among performers and other local residents, curtailing his full immersion in several of the destinations he studied (see, e.g., E. Bruner 1993b; 1994; 1996a; 2001). Yet we can also identify opportunities afforded by his emphasis on the specific, the immediate, the local, the felt, and the intertwining of narrative, experience, emotion, and subjective meaning: these are aspects of human social life that are difficult to grasp from a primary focus on power, politics, and history. Because it begins with fine-grained attention to specific moments of encounter and an empathetic engagement with diverse participants' experiences, Bruner's work stands as an influential model for how anthropologists and other qualitative social scientists can address tourism as a rich field of cultural production in the contemporary world.

This Volume: An Itinerary

This book has two tasks, one general, the other specific. First, at the most general level, the volume offers an in-depth exploration of key concepts, theoretical debates, and methodological concerns that characterize a major strand of ethnographic tourism research today: the experience-near, interpretive-humanistic approach widely associated with Bruner's body of work. Second, individually and collectively, the chapters in this volume—each written purposefully for inclusion here—provide a rigorous assessment of how these concepts and theories take shape specifically in Bruner's oeuvre, how they relate to broader trends in tourism ethnography, and how they can be challenged and extended through analysis of the authors' own original ethnographic material.[17] To assist the reader in using the book as a resource, each chapter carries one or more keywords highlighting its relevance to broader topics and approaches in tourism studies.

The volume is divided into five parts. Part I, "Orientation," contains this introductory chapter. Part II, "Points of Departure," situates Bruner's tourism work in relation to his larger intellectual career and to anthropology as a whole. Chapter 1, "Always in Process: Edward Bruner, American Anthropology, and the Study of Tourism" (keyword: *Formation*), examines Bruner's stature as both a pioneer and a product of broader anthropological trends that ultimately paved the way for ethnographic tourism studies. Written by Nelson Graburn and Naomi Leite, the chapter tracks how Bruner's graduate training at the University of Chicago in the 1950s, his early research on social change and individual creativity, and his subsequent engagement with performance theory and lived experience together paved the way for his eventual turn to tourism research. In particular, Graburn and Leite identify antecedents and elaborations of themes that appear throughout Bruner's work and widely in the ethnographic tourism literature—among them the arrangement of touristic social space, the cultural and contextual basis of meaning, and the centrality of narrative and imagination in shaping tourist experience.

With chapter 2, "On the Emergence of Identity and Borderzones as Key Concepts" (*Genealogies*), Kathleen Adams traces the continual interplay of three major foci in Bruner's work since the 1950s: mobility and migration, cultural encounters, and social identity. While each of these foci is a classic topic in anthropology, Adams demonstrates that Bruner's late-career emphasis on identity dynamics in the touristic borderzone is a direct outgrowth of his earlier work on the mutual entanglement of identity, mobility, and intergroup encounters. In turn, she suggests, this intellectual genealogy can be traced through subsequent generations of scholars, whose work takes the interplay of these three foci into new subject areas within the broad arena of the ethnography of tourism.

Considerations of influence continue in Julia Harrison's chapter 3, "'So in Effect I Was Studying Myself': Knowing (Our) Tourist Stories" (*Influence*). In tracing her own story of entering and pursuing field research on tourism, Harrison sketches how anthropology as a whole was shaken by the intellectual upheavals of the 1980s and 1990s, with Bruner a leading participant. She emphasizes the importance of narrative—the stories people tell one another, the stories they recount to anthropologists, and the stories anthropologists tell to and about them—for ethnographic research, interpretation, and representation. She concludes that Bruner's work directly and indirectly gave rise to themes that now resonate throughout the anthropological ethnography of tourism: the essential distinction between reality "as lived, as experienced, and as told" (E. Bruner 2005b, 19); the role of emotion and affect in shaping tourists' experiences and their subsequent memories of travel; and the analytical importance of reflexivity about the anthropologist's role—both personal and professional—in the tourist encounter.

The book's third part, "Explorations: Concepts and Debates," contains chapters centering on theoretical concepts and approaches that lie at the heart of Bruner's contributions to tourism research. In chapter 4, "'Whatever We Weave Is Authentic': Coproducing Authenticity in Guatemalan Tourism Textile Markets" (*Authenticity*), Walter Little revisits the perennial debate over authenticity in tourism studies, including Bruner's contributions, and turns the question of whether tourists are seeking "authenticity" on its head. He approaches the problem not from the perspective of tourist consumers, nor that of local producers who create "staged authenticity," but instead via an analysis of the mutual engagement of makers of tourist arts and those who buy them in negotiating very different ideas of what the term *authenticity* means. His closely observed, extended accounts of interactions between tourists and Maya textile weavers offer a tantalizing glimpse of the coproduction of authenticity, as he documents how cross-cultural dialogue gradually shifts participants' understandings of what constitutes an "authentic Maya weaving."

Valerio Simoni's chapter 5, "Living in and Reaching beyond the Touristic Borderzone: A View from Cuba" (*Borderzone*), illustrates how Bruner's concept has helped to turn ethnographic work on tourism away from the static hosts-and-guests framework to an approach that foregrounds tensions between discourses, narratives, and imaginaries at work in the encounter. Drawing on his long-term research on interactions between foreign tourists and Cubans in Havana and beyond, Simoni argues for greater attention to linkages between touristic borderzones and other realms of sociocultural creativity, (im)mobility, and personal aspirations. As he demonstrates, for some of his Cuban interlocutors the touristic borderzone is not at all separate from the "everyday life" they return to at the end of the day, as Bruner would have it, but rather provides a more valued social space to which they direct their energies and in which they prefer to dwell. Indeed, via their interactions with tourists, the borderzone offers them possibilities for self-reinvention and even the promise of a much-desired *new* everyday—to be achieved by the transformation of their interactions in the borderzone from "performance time" to "real life."

With chapter 6, "'I Can Feel Them Now, Even as I Write': Hiking Yosemite Falls with the Emergent Subjects of Tourism" (*Constructivism*), Sally

Ness shifts the focus from key concepts to the overall theoretical underpinnings of Bruner's work. Given his central preoccupation with how meaning is made and worlds created in touristic environments, she asks: What are the constituent elements of Bruner's hallmark brand of constructivism, and what is its relationship to phenomenological, social-constructionist, and semiotic perspectives on experience and meaning-making? Taking seriously the idea that the ethnographer's own experiences can be a rich data source, she approaches this question through an auto-ethnographic account of how her consciousness of self and surroundings shifted while hiking an arduous trail in California's Yosemite National Park. Using her narrative as both scaffolding and substance for analysis, she examines classic social constructionism's opposition between first-world (material, environmental, precognitive) and second-world (cultural, linguistic, conceptual) realities, arguing that her experience on the trail suggests that the two may not be as separate as they seem. This, she contends, aligns neatly with Bruner's "pragmatic constructionism," as she calls it, because the latter emphasizes an embodied, phenomenological, always-emergent basis for meaning, one that is not determined by language but rooted in embodied and emotional aspects of any given encounter.

Part IV, "Further Afield: New Departures," presents chapters combining multiple themes introduced earlier in the volume and extending them in novel directions. Building upon the work of Bruner and others in the interpretive-humanistic ethnography of tourism, as well as a wide range of scholars in other traditions, they demonstrate myriad ways that tourism is embedded in and contributes to a vast range of domains of human activity—economic, political, emotional, spiritual, and more. Chapter 7, Kathleen Adams's "'Being a Tourist in My (Own) Home': Negotiating Identity between Tourism and Migration in Indonesia" (*Identity, Mobility, Embodiment*), draws inspiration from Bruner's explorations of Batak urban migrants' use of their cultural pavilion at Jakarta's Beautiful Indonesia in Miniature Park (Taman Mini) to connect with aspects of their own identity. Turning her lens to Indonesian migrants returning to their homeland for visits and vacations, Adams problematizes the classic divides between such categories as returning migrant and tourist, kin-visitor and (part-time) resident. She foregrounds visitors' embodied experiences, as well as the ambivalence, attraction, and revulsion many experience, while embedding her analysis in broader economic and political issues (among them, how homeland tourists' class and rank shape their experiences, as well as the role of government campaigns to harness returning migrant visits for political-economic gain). Adams's chapter thus connects with and extends Bruner's concepts of heritage tourism, identity, embodiment, mobility, and contested sites, integrating them into broader realms of political, economic, and social significance.

In chapter 8, "Beyond Dialogue: Hospitality and the Transformation of Self in Southwestern Madagascar" (*The Self, Narrative, Borderzone*), David Picard turns our attention back to the creative potential of encounters in the borderzone, here in light of Bruner's abiding interest in self-formation as a dialogical process. Picard's ethnographic focus is a Madagascar fishing village, where interactions with foreigners—tourists, missionaries, medics, conservation scientists, even spirits—are embedded in local strategies of hospitality as a source of individual and collective empowerment and well-being. Through cases of local figures whose interactions with foreigners have proved transformational, he shows that each individual understood these outsiders within the framework of master narratives of self and other, in which social relations with spirits, the transmission of qualities and substances, and the containment and channeling of power are brought to the fore. Thus Bruner's concepts of overarching narratives, constructivism, the emergent and dialogical self, and the borderzone are, in Picard's chapter, fruitfully brought into an exploration of local engagements with and experiences of witchcraft, religion, diagnosis and healing, and, above all, socioeconomic change.

The final chapter in this part, Quetzil Castañeda's "'Ideologies at War' at Chichén Itzá: An Ethnography of a Tourism Destination" (*Contested Sites, Identity, Stories*), revisits Bruner's notion of contested sites to question its distinctiveness. Noting that a variety of other approaches to the study of place and space emerged in the 1980s and 1990s, all equally pertaining to contestation over meaning, Castañeda builds on and yet departs from and radicalizes Bruner's analytical framework of Bakhtinian dialogism, performance, and storytelling in an analysis of the site of Chichén Itzá. He recounts a series of episodes that demonstrate the interweaving of economic meanings and cultural values ascribed to Chichén—as highlighted in events relating to the main pyramid's selection for a new list of Seven Wonders of the World, the site's expropriation by the state government of Yucatán, and "invasions" of the site by New Agers, handicraft vendors, and even Subcomandante Marcos—and shows that local identities, political-economic relations, and the stories we tell about people and place are inevitably mutually constitutive. In the process, he identifies and highlights contrasting elements between Bruner's approach and his own.

Finally, part V, "Homecomings and Future Directions," addresses the social history and possible future contours of the ethnography of tourism, as well as Bruner's own intellectual history. Chapter 10, "Taking Tourism Seriously," presents a lively interview with Bruner, edited and with commentary by Michael Di Giovine. This is followed in chapter 11, "Reflections" with a brief response to this volume by Bruner himself. The book concludes with an afterword by Mary Mostafanezhad and Margaret Byrne Swain. Their chapter, "Beyond Anthropology: Ethnography in Tourism Studies," examines how Bruner's concepts and methods—and, more broadly, those of the anthropological ethnography of tourism—have been taken up, challenged, critiqued, and extended by scholars far outside the bounds of anthropology as an academic discipline.

NOTES

1. See, for example, Graburn (1983a); Crick (1989); Stronza (2001); Jamal and Robinson (2009); Leite and Graburn (2009); Stasch (2017); Simoni (forthcoming).

2. There are, of course, exceptions to this rule. By the 1950s, a growing number of anthropologists had begun to examine the interplay of local, regional, and international forces in shaping life for the people studied, including colonialism, industrialization, and labor migration. For an early, influential example, see Gluckman (1940).

3. Van den Berghe presented his lecture in one installment of a faculty seminar on tourism and ethnicity sponsored by the Jackson School of International Studies at the University of Washington. Many of the papers from the seminar were subsequently published in a special issue of *Annals of Tourism Research* (van den Berghe and Keyes 1984). For more on the seminar, see K. Adams (2016). For detailed first-person accounts of this early generation's entry into tourism ethnography, including one contributed by van den Berghe, see Nash (2007).

4. For further examples of ethnographic essays and programmatic statements on the anthropology of tourism from the 1970s and 1980s, see Graburn (1976; 1983a; 1983b); Nash (1981); Crick (1989).

5. Another chapter in the volume focuses on tourist–toured encounters (Loeb 1977), but it is based primarily on observations in markets where the ethnographer was already doing long-term research with the local population.

6. In a related strategy, for his research on Japanese tourists Graburn (1983c) built on background knowledge gleaned during repeated visits to and leisure travel alongside his wife's extended family in Japan (see Nash 2007, 103).

7. It is important to note that there is more than one mode of reflexivity in anthropology. The term encompasses a range of meanings, from an approach to collecting data in fieldwork to a framework for analyzing ethnographic evidence, a style of ethnographic writing, and a principle by which to formulate research problems (Castañeda 2008). For a reflexive exploration of the relationship between practices of note-taking in the field, the resulting corpus known as field notes, and the published ethnographic text, see Sanjek (1990).

8. Concurrent with this efflorescence of popular imagery, qualitative social scientists began to analyze touristic representations of formerly colonized peoples, including travel literature (K. Adams 1991; Castañeda 1996; Hutt 1996), postcards (Aloula 1986; Edwards 1996), and tourist brochures and advertising (K. Adams 1984; Moeran 1983; Thurot and Thurot 1983; Dann 1996).

9. Of course, Bruner's approach to the anthropology of tourism was not the only one on offer, and from the 1990s to the present anthropological ethnographers of tourism have worked in a wide range of other theoretical and methodological traditions. For example, applied, development, and environmental approaches have proliferated, but are not discussed here as they address a different set of issues. For overviews, see Nash (1996); Stronza (2001); Wallace (2005); and Chambers (2009).

10. At the time, theoretical factionalism was raging in anthropology, primarily between neopositivists (environmental determinists, including sociobiologists) and more humanistic scholars (interpretive and symbolic anthropologists). Bruner and Judith Becker neatly summarized the latter perspective at the time: "It is this domain of expressive culture, the domain par excellence of creativity and imagination, that is most revealing about the essence of society. It is in the realm of pure fantasy rather than in the ecologically determined that one can penetrate most deeply into the life of a people" (Bruner and Becker 1979, 9).

11. These roles include president of the American Ethnological Society (1980–1982), president of the Society for Humanistic Anthropology (1987–1991), and member of the board of directors of the American Anthropological Association (1989–1991).

12. This constructivist approach to the variable meaning and experience of sites has been taken up by a number of scholars and applied to considerably more complex cases of narrative, embodiment, and experience in historical sites, including those where the destination is characterized by absence. See, e.g., DeLyser (2001); Chronis (2005; 2012); and Leite (2007).

13. For discussions of the concept of encounter in interdisciplinary tourism studies, see Crouch (1999); Crouch, Aronsson, and Wahlström (2001); Gibson (2010); and Mostafanezhad and Hannam (2014, 6).

14. A particularly clear example of this approach is "Tourism in the Balinese Borderzone" (1996b), Bruner's multilayered exploration of crosscutting expectations, interests, and understandings generated in an unforgettable incident involving an American package tour group, Balinese dance performers, a fellow anthropologist, and Bruner himself as tour guide.

15. For further discussion of ethnographic methods in anthropological research on tourism, see Graburn (2002); Frohlick and Harrison (2008); Simoni and McCabe (2008); K. Adams (2012); and Andrews, Jimura, and Dixon (2019).

16. For similar uses of the concept of narrative in tourism ethnography, see Dahles (1996); Selwyn (1996); Ebron (1999); Harrison (2003; 2013); Chronis (2005; 2012); Leite (2017); and Picard and Zuev (2014).

17. As is standard in ethnographic writing, contributors to this volume have removed identifying factors and given pseudonyms to their respondents, unless explicitly granted permission otherwise.

Part II

POINTS OF DEPARTURE

Chapter One

{Formation}

Always in Process

Edward Bruner, American Anthropology, and the Study of Tourism

Nelson Graburn and Naomi M. Leite

> In all of my research since the late 1940s there has been a continuity of focus on encounters, change, mobility, and process. In my dissertation (1954) I studied Native Americans: why and how they changed, why some individuals changed more than others, who left the reservation for the city, and which aspects of culture were more persistent. In 1957 I began fieldwork in an Indonesian village among the Toba Batak of Sumatra, studying kinship, change, and urbanization in the city of Medan. . . . For me cultures were never static, enclosed, integrated entities but were always in process; we all enter society in the middle.[1]

Edward Bruner has been a major contributor to the study of tourism for nearly three decades. Known in the 1950s for his original work on American Indian kinship and social change, and since the 1960s for research on urbanization and identity in Indonesia, he later became a leading figure in the anthropology of symbolism, ritual, and experience (e.g., E. Bruner 1984b; Turner and Bruner 1986). It is from the latter set of interests and from his travels as an expert-lecturer tour guide that he launched into tourism studies, and here his work took a profoundly reflexive turn. He paid particular attention to the experience of the anthropologist, as well as the performance of tourists and the peoples and cultures they visited (Graburn 2008).

While the shift from kinship and urbanization to symbolism, tourism, and performance might seem a dramatic one, even before Bruner went to Indonesia in the late 1950s his work was directed at problems of symbolic consistency and inconsistency within cultures and communities undergoing rapid social change. This led him to a larger questioning of the structural functionalist "organic" model of society, to focus on problems of cultural transmission and change. Within this general area, Bruner was always aware of the possible non-coterminous relation between culture and society and, above all, the special position of the individual in each. These are fundamental theoretical consistencies throughout his work, from his early studies of Mandan-Hidatsa kinship and acculturation to his writings five decades later about narrative, the problem of authenticity, and new forms of societal integration brought about by globalization and enacted in touristic borderzones.

In this chapter we show that Bruner's interest in these topics is closely tied to his entry into the discipline at a critical moment. Throughout the history of American anthropology there has been concern with forms of sociocultural change, whether debates over internally evolving versus externally imposed changes; the integrity, fragmentariness, or fluidity of culture; or the roles of individuals and subgroups within community cultures. Nonetheless, the particular timing and locale of Bruner's training are highly significant, and indeed laid the groundwork for his development of a constructivist approach and his eventual focus on tourism.

In making this argument, we draw in part on personal experience: Bruner and Nelson Graburn, coauthor of this chapter, studied at the University of Chicago within a decade of each other—Bruner in the early 1950s and Graburn in the early 1960s. The anthropology they entered was radically different from what it has become today. From years of discussion, they know they shared as initial role models anthropologists who are now rarely mentioned: Ralph Linton, Robert Redfield, Sol Tax, and Fred Eggan. Both began their careers studying Native North Americans—Bruner the Mandan-Hidatsa of the Plains, and Graburn the Inuit (then called Eskimos) of the eastern Canadian Arctic.[2] And while at the beginning of their careers both of them would have thought it a far-fetched topic, both went on to become deeply involved in studying tourism. These similarities did not arise by chance.

AMERICAN ANTHROPOLOGY AT THE UNIVERSITY OF CHICAGO

Let us begin by teasing apart some of the origins of American anthropology as Bruner (and Graburn) first encountered it. The discipline's "ancestors"—Franz Boas, Alfred Kroeber, John Swanton, and the like—had rejected the unilinear cultural evolutionism of the nineteenth century, wherein sociocultural changes in any and all societies were said to follow a uniform path from "simple" to "complex" (Morgan 1877). They sought instead to reconstruct traditional cultures in restricted areas through detailed field research with older informants. They looked at culture, not at people, per se, and argued that culture "areas" could be discerned and "traits" mapped, revealing patterns of diffusion. Because of the devastation of Native American communities, however, these early anthropologists largely had to develop histories of cultural diffusion through interviews and analysis of material culture (Boas 1896; Kroeber 1931). Using these methods, Edward Sapir, a student of Boas, published a brilliant work on change and Native American peoples (1916). He briefly taught at Chicago (1925–1931) and while there influenced those, including Eggan, who would ultimately become Bruner's teachers. Efforts at reconstruction continued with the next generation, who similarly faced reservation-confined communities in the Plains and Southwest and the still-more devastated native populations of the Eastern Woodland and East Coast (e.g., Steward 1950).

Two factors changed the direction of both method and theory in American anthropology. First, two British-trained anthropologists, Bronisław Malinowski (1944) and A. R. Radcliffe-Brown (1952), fathers of anthropological functionalism, each spent several years lecturing at a major American university. Radcliffe-Brown taught at Chicago from 1931 to 1937, during which time both Eggan and Tax—later to become Bruner's teachers—were doctoral students there;[3] Malinowski taught at Yale from 1937 to 1942. Having conducted long-term ethnographic research in "intact" communities in the far-flung Empire, they disdained the aim of what they called the Boasians' "conjectural history," conducted for its own sake, though they were willing to entertain some examination of history and change. Both men traveled and lectured widely during their time in the United States and fundamentally influenced the direction American anthropology would take. But it was Radcliffe-Brown's residence in Chicago, together with the important contributions of Robert Park, Robert Redfield, and others, that made the Department of Anthropology at Chicago such a seminal place for the next two decades (Stocking 1979).

By the mid-1930s, American ethnographers could no longer claim they had access to the "pure traditions" of Native cultures—even the oldest of their informants could not remember precontact times—nor could they support the vision of Euro-American culture as superior and hence inevitably dominant in processes of culture change. A small group spearheaded by Robert Redfield, who had trained at Chicago and joined the faculty in 1927, developed a new paradigm known as acculturation. Announced in an extraordinary publication coauthored by Redfield, Linton, and Melville Herskovits, the "Memorandum for the Study of Acculturation" served as a manifesto and international call for participation in the study of "those phenomena which result when groups of individuals having different cultures come into continuous first-hand contact, with subsequent changes in the cultural patterns of either or both groups" (Herskovits 1938, 10). Released near-simultaneously on three continents—in

Man (UK, 1935), *Oceania* (Australia, 1935), *American Anthropologist* (US, 1936), and *Africa* (UK, 1936)[4]—their argument shifted the prevailing approach to social change away from evolution, diffusion, and conjectural history and toward a vision of cultures in contact.

Eggan (1954), in particular, masterfully combined the American historical and British functional research traditions by examining how Native American cultures were changed through displacement and diverse forms of contact—phenomena that were present all along but previously ignored in the search for the "pure traditions." In addition, by the early 1950s the University of Chicago had become the spearhead of anthropological studies in Latin America under Redfield (1941), followed by Tax (1953). Fieldwork in this new locale confronted anthropologists with vast areas of *mestizo* or "peasant" populations who had lived under direct rule of Spain and the Catholic Church for centuries. Redfield (1941) proposed a rural-urban contact continuum of the "little" or "folk" tradition and the "great" or "national" tradition, arguing that the village and the nation had been in continual contact for hundreds of years and were thus impossible to fit into simple schemes of human cultural evolution.

Although the "Memorandum," written by senior anthropologists of Latin America, Africa, and Native North America, and published in the major journals of three continents, gave scholars license to study ongoing change and laid to rest forever the directive that the prime task of the sociocultural anthropologist was to reconstruct all forms of "primitive" societies, it was not satisfactory as a guide for research. The study of acculturation assumed that societies in contact necessarily had an influence on each another, that traits or customs were "borrowed" to differing degrees. When one society was dominant, as in the case of Euro-Americans in North America and Europeans in sub-Saharan Africa, or urbanites over "folk" (usually rural indigenous peoples), as in Latin America, this borrowing was assumed to be rather one-sided.

But there was no overall theoretical model of precisely how acculturation happened. Malinowski (1945), for example, posited that in colonial Africa there would be an accommodation between white and native societies, such that matching institutions[5] on each side would accommodate and produce a new reality: law on law, art on art. His predictions proved groundless.[6] Tax, on the other hand, rebelled against the notion of acculturation as a slow but inevitable process in which the dominant culture influenced the others. He is alleged to have said that acculturation simply does not occur and to have predicted that there would be Indian enclaves in North America for the next thousand years; yet it was evidently a chance personal experience, rather than his own research, that brought him to this conclusion. In 1952 he went on a tour to Algeria. There he met some local Jews and, being Jewish himself, could not help comparing them with the American Jewish culture he knew. Although there were obvious cultural differences, he wrote then that he found a "great similarity of the mental make-up of the Jews here with those at home. . . . They were essentially the same" (Morris-Reich 2008, 47nn89–91).[7] This was not a statement of biological determinism, of course, but an argument for the uneven and unpredictable character of cultural change overlooked in the prevailing acculturation model.

Despite the lack of consensus, the growing attention these early studies gave to historically contingent, localized, and multiscalar forces of social transformation prompted many anthropologists to become activists, intervening in the courts and in Washington on behalf of the people they studied. Tax started a movement at Chicago that he called "action anthropology," which sent students to engage and organize on behalf of the communities they studied (Tax 1952). This was no longer anthropology as the search for vanishing traditions or even for smoothly functioning culturally homogenous communities. For Tax and his colleagues, the communities studied by anthropologists were made up of individual people with present-day problems, not all in agreement, and very much affected by the vast world around them. Taken together, these innovations paved the way for a new generation to study the world's fragmenting, assimilating, adapting, urbanizing societies as a primary focus.

BRUNER, DOCTORAL STUDENT

When Bruner arrived to begin doctoral study at Chicago in 1950, he took courses with Redfield, Eggan, and Tax.[8] He carried out his doctoral fieldwork on the Fort Berthold Reservation in North Dakota. Fort Berthold was home to settled descendants of the Mandan, Hidatsa, and Arikara Plains Indian tribes who had lived in the region for hundreds of years. From 1947 to 1954 approximately 80 percent of the reservation's residents were relocated to make way for a federal hydroelectric dam project, the Garrison Dam. The tribes' agricultural and grazing lands were flooded to create the dam's reservoir. The massive loss of livelihood and destruction of traditional lifeways attracted the attention of sympathetic groups, among them the American Missionary Society, which arranged for a grant to the University of Chicago for anthropological study of the difficulties caused by forced resettlement. This was the Fort Berthold Action Anthropology Project, headed by Tax and conducted by Chicago anthropology students under his supervision.[9] The students could pursue whatever topic they wanted, provided that they sent copies of their field notes to Tax (E. Bruner 2012, 139).

Bruner spent his first summer of graduate school at Fort Berthold, studying social and cultural change. Like Tax, he immediately noted problems with some of the assumptions behind the study of acculturation:

> The thrust of my research was to show that although there was change, complete assimilation was not happening; indeed, Indians had retained a distinctive language and set of cultural practices. Where change had occurred, I wanted to uncover the mechanisms involved. . . . What intrigued me was to discover that reservation culture was not an "acculturated" mix of "Indian" and "white," as if it were an amalgam that had emerged from a Cuisinart blender, but that members of the community learned two sets of cultural norms, one Indian and the other white, and they practiced one or the other, depending on the situation. (E. Bruner 2012, 141)

In exploring community members' choices about which cultural system to draw on and enact in different social settings, Bruner was as much concerned with the individual as with the village or the tribe. At that time, the leading edge of American anthropology was psychological anthropology, then called "culture and personality" (e.g., Hallowell 1947; Mead 1935); rather than focusing on "cultures" as unified wholes, scholars looked at how individual life experiences, the family, child rearing, and kin relations made people who they were. Bruner was especially interested in how the individual was integrated into the group and the family of orientation, and how those groups and families integrated or differentiated themselves with respect to the larger, socially complex, multicultural society of the United States.

His graduate fieldwork followed in Eggan's footsteps (1933; 1937a; 1937b), examining how Mandan-Hidatsa kinship terminologies, ideologies, and behaviors changed from traditional, lineal systems to resemble the bilateral system of the majority of "white" Americans. But unlike Eggan, he found no gradient of acculturation with intermediary types; instead, "the Indian way" and "the white way" coexisted on the bilingual reservation. After the tremendously disruptive series of events surrounding their relocation, he learned, some Mandan-Hidatsa had decided to discard the traditional social organization, with its obligations of lineage and moiety ties, and had adopted the prevailing bilateral nuclear family system, with its emphasis on individual and nuclear-family accumulation and advancement. Although it was prompted by culture contact, this was neither passive nor inevitable acculturation: only some Indian families adopted the new kinship model, consciously shifting toward "white" ways and encouraging their children to "become like whites" for the sake of their future (see also Adams, chapter 2, this volume).[10] In other words, Mandan-Hidatsa individuals weighed their options and exercised agency in making their way through life (E. Bruner 1954; 1955; 1956b; 2012).

Broadly speaking, Bruner's (1954) dissertation carried the seeds of two themes that appear throughout his work, both involving individual responses to the community's relationship with and integration into a larger social system. First, while Mandan-Hidatsa families may have been internally differentiated by language, cultural practices, and kin-term usage—"Indian" or "white"—all were undergoing legal and social integration into American society, whether or not they adopted "white" ways. Years later, when studying tourism in Bali, Bruner described a similar situation: some Balinese directly engaged with the tourism sector and others did not, but the latter did not represent untouched enclaves of "authentic" Balinese culture. The issue was not "the 'impact' of tourism on the culture, as if tourism were like a meteor hurtling in from the sky," but instead the ongoing coproduction of a society in which tourism was already embedded (E. Bruner 2012, 152).

The second, related theme is that of individual experience, decision-making, and resulting performance. The special form of acculturation that Bruner found was the incorporation of the new "American" culture (categories, norms, language) alongside the persistent traditional "Indian" culture. While some individuals chose to live within their traditional culture on the reservation, others had chosen "American" culture, spoke only English, and moved off the reservation. But a significant proportion of reservation residents were fully "fluent" in both cultures and, like bilingual people everywhere, chose situationally which set of norms to apply. This led Bruner to develop a more constructivist view of kinship, far ahead of his time. In structural functional social anthropology, kinship systems were seen as more formal and mechanical, and there were few if any analyses of how individual people constructed their kinship roles. Yet Bruner found significant variation from one situation to the next. One case he cites, for example, involved a young girl and an older woman who were icawi ("father's sister") to each other in the Mandan-Hidatsa kinship system. Although teasing was not considered appropriate in that relationship, he heard the girl openly tease the woman, who replied, "You're supposed to be my icawi . . . and you are not supposed to tease me, but that's O.K., we can take it the 'white' way" (E. Bruner 1955, 848). People had choices, and they constructed their kinship roles differently depending on the specific situation at hand.

Bruner's publications on this material (1955; 1956b) mark a turning point in the discipline's move from a positivist to a constructivist approach in studying kinship. Although it was not immediately apparent, the influence of his early mentors, Chicago training, and initial fieldwork experiences had paved the way for this shift, just as they did for Graburn some years later, when he encountered similar predicaments in the construction of kinship roles among Inuit adults and children.[11] Among the many anthropologists struggling with the results of more intensive field research, which brought up individual differences and choices in situations of social and cultural change, Bruner was a pioneer.

BRUNER'S JOURNEY—FROM TOBA BATAK TO TOURISM

Few anthropologists today are aware of Bruner's Native American research; his work in Indonesia is much more widely known. Yet there are continuities between his first and second major anthropological explorations. In 1957 he traveled with his wife to Sumatra, where they initially settled in a rural Toba Batak village. His plan was to investigate kinship, still a central topic in sociocultural anthropology, through intensive participant observation. However, civil unrest and armed insurrection made it too dangerous, and they had to relocate to the city of Medan for safety (E. Bruner 2012). Here Bruner encountered another instance of Redfield's rural-urban continuum: whereas the urbanized Mandan-Hidatsa shunned their more traditional neighbors, and many had left the reservation altogether, the Batak valued their ethnic links, maintaining their contacts through regular visits from the city to the village and vice versa (see Adams, chapter 2, this volume).

After settling at the University of Illinois in 1962, Bruner focused at first on urbanization and ethnicity, but increasingly drew on the symbolic interactionism he had encountered as a graduate student (2005b, 17; 2012, 142; see also Di Giovine, this volume), moving farther from the positivism of anthropology as he first encountered it to investigate the construction of meaning from experience. Taking what he would later call an "inner perspective"—trying "to understand the world as seen by the 'experiencing subject'" (E. Bruner 1986c, 9)—he analyzed situations of social change among the Toba Batak using native categories, but emphasized that the categories themselves were flexible: "There is no such thing [for the Batak] as a modern or traditional role. . . . The occupant of a role devises a performance in reference to all the significant others in the context. As [Ralph H.] Turner . . . has suggested, it is role making, not role taking" (E. Bruner 1973, 225). Here we see his use of performance as a central element in the creation of meaning, a concept he had initially encountered in ongoing dialogue with Milton Singer in the 1950s (E. Bruner 2012, 142).

Less explicit but equally present is his predilection for American psychological anthropology, which informed his work particularly in the 1960s and early 1970s (e.g. E. Bruner 1964); he argued, for example, that the lack of psychological conflict he observed in response to the multiplication of social roles could be confirmed by projective tests (the Thematic Apperception Test and the Rorschach) that he had administered to thirty-two rural and urban individuals. He also tried out another directions—the Barthian semiotic, for example, in his research and interpretation of gender differences in graffiti on bathroom walls (Bruner and Kelso 1980). Here he was treading on the same theoretical ground as a colleague who was soon to become significant in his work, Dean MacCannell (1976).

The study of the Toba Batak had a lasting effect on Bruner's theoretical directions, especially regarding the importance of ritual and play in their lives, which translated into performance in his analyses. Again, in this he was influenced by Singer's work on the concept of cultural performance (1972), but also Victor Turner's studies of performances in ritual and pilgrimage (1969; 1974) and, later, the work of Turner's colleague Richard Schechner (1985), a performance theorist.[12] From that time on, Bruner emphasized ritual, creativity, and play, as in his edited volume *Text, Play, and Story* (1984b), focusing on settings in which rules are loosened and participants can be imaginatively and socially creative. In his collaboration with Turner, who was then in his final years, they focused on culture as experienced—how cultural performance was registered and remembered. A key session at the 1980 American Anthropological Association meetings on "the anthropology of experience" consolidated their efforts, with the papers published in a coedited volume by the same title (Turner and Bruner 1986).

Clifford Geertz, also a participant in the conference, wrote the epilogue to *The Anthropology of Experience*. Geertz, who had arrived at Chicago soon after Bruner left, was an obvious fellow traveler, a major anthropologist of Indonesia, an antifunctionalist, and an antipositivist, whose interpretive anthropology focused not only on the creation of meaning (Geertz 1973), but even more on the co-creation of meaning between the anthropologist and local interlocutor. Yet, Geertz is rarely mentioned in Bruner's work, even though we know from the few quotes he cites that they agreed on a wide range of things—for example, on the illusory nature of the distinction between "authentic original" and "fake copy" (E. Bruner 1994, 407), when all are creations sui generis; on the "museum view of heritage," which tries to freeze living traditions in the name of authenticity (E. Bruner 2005d, 225); and on Taman Mini (an ethnic theme park in Jakarta) as a model not *of* the nation but *for* it (2005d, 230; cf. Geertz 1966). In the latter case, Bruner (2005d, 230) explains that Geertz's view is "top down," whereas his is "from below, from the inside of the house," enabling him to reveal that Batak visitors experienced the theme park not as an image or text, as Geertz would have it, but as life.

During this same intellectual period another fellow traveler, MacCannell, also sought the meanings of people's interactions as if they were performers on a stage (1976), following Erving Goffman (1959),[13] who had followed Shakespeare.[14] Goffman's view of creative role-playing by actors on a public stage superseded the more structural-functional view of actors (Merton 1949), which assumed that specific roles entailed culturally predetermined behavioral expectations. Reading Goffman, one can see an actor playing a role on a social stage to impress someone or to show obedience or to wheedle a gift, leaving the slight impression that the behavior is forced or deceiving, which raises the question: What is (left of) the "real" person off stage? Many readers answer that if "staged authenticity" is present or enacted on stage, then "real" authenticity must be found in the back stage, and that, indeed, is what MacCannell (1976) argues that tourists are seeking—a point that Bruner has vigorously contested (see Little, this volume).

In this intense period of Bruner's intellectual career, two further events came together that put in place the last two major factors that allowed him to emerge abruptly on the scene as a full-fledged, sophisticated anthropologist of tourism. In 1982–1983, he took a sabbatical in situ at the Center for Advanced Study, University of Illinois, and threw himself deeply into the literature he had been exploring for several years in the faculty reading group known as the "Unit for Criticism and Interpretive Theory" (E. Bruner 1986a; 2012, 149). With the group's members he read Derrida, Barthes, Dilthey, Bakhtin, and Habermas, among other authors. He developed his approach to narrativity from Bakhtin (1981) and published his first analysis of a tourist site, "Dialogic Narration and the Paradoxes of Masada" (Bruner and Gorfain 1984). He also strengthened the underpinnings of his processual, constructivist approach, wherein culture is always emergent—that is, generated as it is expressed—and derives meaning from its sociocultural, linguistic, and temporal context. From such a perspective everything is "new" (and equally "authentic"), even if it is a copy of something else, and meaning can never be repeated exactly because it always resides within temporal and contextual flows. As Bruner wrote, "I reject concepts such as simulacra that are so privileged by some postmodern scholars . . . in part because what is rejected is a new culture constructed specifically for a tourist audience. There is no simulacrum because there is no original" (2005b, 5).

The second and perhaps most important factor was Bruner's engagement in a study abroad program that he led with performance studies scholar Barbara Kirshenblatt-Gimblett, a professor at New York University. For the 1983–1984 school year, they took a group of students to Japan, Bali, India, Thailand, Kenya, Egypt, and Israel (see Di Giovine, this volume).[15] During the eight-month program Kirshenblatt-Gimblett shared with him the subtleties of performance as understood in her field, and they applied the concept to the many tourist attractions they visited during their trip. They subsequently published several coauthored papers, including the widely cited "Maasai on the Lawn" (Bruner and Kirshenblatt-Gimblett 1994; see also Bruner and Kirshenblatt-Gimblett 1987; 1989).

Ever the ethnographer, Bruner also kept an eye on the traveling unit. In his role as a guide to young Americans in foreign countries in which he was not particularly expert (unlike his excursions as a guide in Indonesia at later dates), he must have become aware of the meaningfulness of the experience for his student-tourists and, on a different plane, his own attempts to make and transmit meaning as a guide-tourist. He must have studied all the printed and visual materials, as well as other components forming participants' expectations, or "pre-tour narratives," and watched as the experiences were sorted and built into re-formed narratives during the tour, then reformulated into posttour narratives that were passed on to others, along with mnemonic devices such as photos, movies, souvenirs, clothes, and possibly even foods. He does recount the post-tour gatherings of wealthy Americans after their Indonesian trips (see, for example, 2005b), but he has written far less about the impact of that early study abroad trip on his thinking.

From these crucial experiences, he began to see the "delicious ambiguity" (2005b, 1) of being simultaneously a tourist, a guide, a teacher, and an ethnographer. This brought to the fore the exploratory reflexivity, itself a form of dialogic narrativity, that had always been part of his personal and intellectual approach. At that very time reflexivity had become a central challenge in American anthropology—as, for example, in James Clifford and George Marcus's influential edited volume, *Writing Culture* (1986), which urged ethnographers to reflect critically on the nature of their selection, organization, interpretation, and explanation of "other peoples' cultures" in ways very similar to Bruner's work.[16]

Although Bruner rarely entered directly into the fray of theoretical debates in interdisciplinary tourism studies—"authenticity" being a notable exception—or even among other anthropologists focusing on tourism-related matters, several of his key themes are woven throughout the literature. One of these is the arrangement of (touristic) social space, a concept based on Goffman's dramaturgical "stage" (1959), the public social space in which we enact our social lives and make ourselves up as we go along, in interaction with other people. This simple model implies that all social actors are either on- or off-stage, depending on whether they are interacting with others. MacCannell's "stage," in his initial analysis (1976), is a physical and conceptual space specifically of performance, to which local people come out from their private "backstage" lives to engage with tourists; tourists, too, come to the encounter from somewhere else, but contrary to Goffman's model, they are not similarly "on stage."

In Turner's thinking, a "social drama" (1969; 1974) is a temporal space. It takes place within rituals and, according to his later work, during pilgrimages and tours. Turner draws on Arnold Van Gennep's idea of liminality (1909) and imbues it with a temporary *communitas*, a social state of equality and minimal interhuman barriers allowing novel thoughts and creativity, which he calls "anti-structure." In MacCannell's subsequent work, touristic space becomes an "empty meeting grounds" (1992), empty of the normal social rules—much as in liminality—and fostering creativity and emergent identities. This theatrical space (Edensor 2001; Schechner 1985) shares much in common with Johan Huizinga's (1950) space and time of "play," which occurs away from home at a special time. Bruner's borderzone (1996b; 1999; 2005b) is a culmination of these earlier concepts, conceived as the meeting grounds of two distinct cultural communities for different but mutually satisfactory purposes, from which each could step away order in to resume their "normal" identities at home (see also Simoni, this volume).[17]

Just as the concept of the liminal, creative social space of touristic encounter has had many sources, so too does the concept of cultural meaning systems that lead individuals (such as tourists) to experience their interactions in particular ways. This is, of course, a central component of social constructionism, a theoretical perspective with which Bruner's work is strongly identified (see Introduction and Ness, this volume), but it also has a very long history in anthropology, initially in the study of "worldview," which Redfield (1952) identifies as a shared conceptual system containing notions of self and other, time and space, and the sacred and the profane (Kearney 1975; 1984). This model rests on the idea that enculturation inculcates implicit assumptions about the world and one's place within it, as well as conceptual tools with which to interpret what is going on in diverse settings.

Gregory Bateson's (1972) concept of "frames" pinpoints the culturally specified, basic units of interpretive context that enable people to understand a given set of behaviors and events as, for example, "play"—or, closer to our purposes, a "tourist dance" (E. Bruner 2005b, 21). Goffman's "frame analysis" (1974) broadens Bateson's concern with contextual framing to encompass all shared "frameworks," or "schemata of interpretation," that members of a social group draw on to organize, interpret, and react appropriately to occurrences and interactions. This approach has clear echoes in Dorothy Holland and Naomi Quinn's later concept of cultural models, "presupposed, taken-for-granted models of the world that are widely shared . . . by members of a society and that play an enormous role in their understanding of that world and their behavior in it" (Holland and Quinn 1987, 4). Bruner's emphasis on the sense-making importance of narrative (1984a; 1993b; 1994; 2005b) rests neatly within this larger interdisciplinary discussion, stressing as it does a culturally inflected conceptual structure that gives meaning to otherwise potentially incoherent experience.

Two forms of narrative that Bruner highlights in relation to tourism—metanarrative and master narrative—warrant particular attention here. Metanarratives, "the largest conceptual frame within which tourism operates"

(E. Bruner 2005b, 21), are not specific to any one location, but function as conceptual schemas for the journey. Thus, for example, the experience of a luxury tour of African wildlife and tribal cultures nests within the larger framework of global colonial encounters between "older, wealthier whites" and "younger, poorer darker-skinned people," in which intrepid travelers are given a last glimpse of "authentic primitive cultures [that] are being eroded by the forces of modernization" (2005b, 21). Like Bateson's frames, Goffman's frameworks, and Holland and Quinn's cultural models, such metanarratives convey generic roles, dynamics, landscapes, relations, and outcomes that give meaning to tourist experience.

"Master narratives," on the other hand, are specific images of people and place—"the African primitive, the Balinese island paradise, Egypt as the land of the pharaohs" (2005b, 22)—promulgated in popular culture, tourism marketing, and other widely circulating systems of imagery, inevitably shaping tourists' expectations and their experiences of the destination. In Tom Selwyn's (1996) formulation, these are tourist "myths"—widespread, idealized images or "stories" of types of places and people, drawn from diverse sources and continuously reproduced by the tourism industry—that serve to connect specific destinations with more general preoccupations and desires in the tourist's own society. In both cases, as Claudia Strauss (2006, 329) writes regarding the closely related concept of "imaginaries," "these are shared, implicit schemas of interpretation, rather than explicit ideologies." The more recent term *tourism imaginaries*, which Noel Salazar (2012, 864) defines as "socially transmitted representational assemblages that interact with people's personal imaginings and are used as world-making and world-shaping devices," reflects the influence of each of these earlier ideas. Formed of images, discourses, fantasies, stereotypes, advertising, folklore, and more, these imaginaries prompt tourist expectations, provide fodder for tour guide presentations, and lend meaning to the traveler's experience (Leite 2014; Salazar and Graburn 2014; Skinner and Theodossopoulos 2011)—all phenomena Bruner has richly described.

CONCLUSION

Throughout his career, Bruner has been central in forging a path in anthropological theory involving agency, creativity, experience, and narrative. But he has also been very exploratory, venturing across disciplines into phenomenology, semiotics, linguistics, and psychology. He has journeyed in parallel with a number of fellow travelers, not all of whose names have shown up in his primary publications, among them sociologists like Goffman, MacCannell, and Ning Wang (1999) and anthropologists such as Turner, Geertz, and Clifford (1988; 1997). In their own writing, only some of which has dealt with tourism, these and other scholars have produced overlapping but not identical formulations of central ideas in Bruner's work, including the arrangement of (touristic) social space, meaning as contextual and always-emergent, and the importance of broadly circulating cultural narratives, or imaginaries, in shaping (tourist) experience. Perhaps we can say that great minds tend to think alike—but their intellectual products are also highly context dependent.

This discussion of Bruner's intellectual trajectory prompts two different sets of conclusions. First, Bruner is simultaneously a pioneer and a product of his intellectual milieu. No scholar emerges sui generis: the theoretical ferment at Chicago in the mid-twentieth century clearly laid the groundwork for Bruner's remarkably long and creative career, as did his close collaboration with a substantial cohort of anthropologists exploring postmodernism in the 1980s—among them James Boon, Barbara Myerhoff, Keith Basso, James Peacock, James Fernandez, Renato Rosaldo, and Barbara Babcock. But he has also been a pioneer in bringing together diverse streams of thought and topical emphases in novel combinations, across disciplinary boundaries.

Second, it could be argued that Bruner's contributions to the anthropology of tourism began long before he turned his attention to tourism—in his theoretical work on play, experience, and its expressions; and in his emphasis on interpretive, constructivist, and reflexive approaches. Yet it was only in the latter half of his career, when he became involved with study abroad and then served as a tour guide himself, that he happened upon tourism as an object of study; in fact, many of his publications on the subject appeared only after his retirement in 1994. To what extent, then, can Bruner be characterized as an anthropologist of tourism? Is he more accurately described as a key early thinker for the subfield, or a relative latecomer? Clearly his work has been foundational for the generations that have followed him—including his writings from the 1950s to the 1970s, far in advance of his tourism work.

To repeat the phrase, there are always delicious ambiguities in how we make sense of individuals, their roles, and their interactions—something Bruner has relished throughout his career. At times he found such ambiguities to be revelatory, as in his dual role of performing as a tour guide and as an ethnographer at the same time; at other times they were uncomfortably awkward, as when he talked to one man in the village of Fort Berthold for two hours without realizing that he was Native American.[18] And these ambiguities could be productive and even joyous, as well. As Bruner (1999; 2012) tells it, because he and his wife had become incorporated into the Toba Batak kinship system during their 1957 fieldwork in a rural village, when they later relocated to an urban environment they were able to continue their research thanks to extended kin ties: their Batak lineage relatives invited them to attend events and incorporated them into the larger community. Here the ambiguity was productive, blurring the role of the anthropologist with that of a kinsman. Forty years later Bruner, his wife, and their adult children traveled to Toba Batak, where they were joyously welcomed back and treated as honored kin. While he may not have set out to become an anthropologist of tourism, tourism scholars, too, embrace the ambiguity of his role and affirm his status as a lineage relative in the tribe.

NOTES

1. From Bruner's biographic statement on the University of Illinois website (since removed), accessed October 1, 2014. https://anthro.illinois.edu/faculty/bruner/. See also Bruner 2005e, 206–7: "As I look back, I experience a continuous Ed Bruner, one anthropological self, not a positivistic traditional period followed by a postmodern hermeneutic humanistic phase."

2. This chapter builds upon Graburn's earlier paper, "Ed Bruner: Before Batak—Kinship and Change," presented at the 2005 annual meeting of the American Anthropological Association.

3. See Eggan 1933; Tax 1935.

4. See Redfield, Linton, and Herskovitz 1935a; 1935b; 1936a; and 1936b.

5. For Malinowski, institutions are human sociocultural structures, found in each society, that satisfy basic human psychophysiological needs, such as law for redress, agriculture and cooking for nutrition, and religion and ritual for allaying fears and uncertainties.

6. Malinowski (1945) based this prediction on his reading of the works of others, mainly his students.

7. Here we would agree with Georges Van Den Abbeele (1980), who asserts that tourists are social scientists of a kind—a supposition Bruner might also endorse.

8. For Bruner's account of his graduate school experience and early career, see Bruner 1999; 2012; and Di Giovine, this volume.

9. See Manuscript 4805, University of Chicago, Fort Berthold Project Records. Sol Tax—Fort Berthold Action Anthropology Project. National Anthropological Archives, Smithsonian Institution. See also Edward M. Bruner Papers, Series 6: Native American Materials, Box 16; Series 7, Box 24: Slides, Fort Berthold Reservation. University Archives, University of Illinois at Urbana-Champaign.

10. Many others had already decided to leave the reservation in order to more fully "assimilate" (Bruner 1953).

11. See Graburn 1964.

12. For Bruner's first-person account of these influences, see Di Giovine, this volume.

13. Goffman was also a doctoral student at Chicago from 1949 to 1953, in sociology, but Bruner (2005b; 2012) makes no mention of having known him at the time.

14. From Shakespeare's *As You Like It*, Act II, Scene II: "All the world's a stage, And all the men and women merely players; They have their exits and their entrances, And one man in his time plays many parts."

15. Until his retrospective writing at the turn of the twenty-first century (1999; 2005b; 2005e; 2010; 2012), Bruner was surprisingly mum on the details of these travels. In fact, even today some of the information is found only in his typescript curriculum vitae. This omission parallels anthropologists' failing to specify the nature of the colonial situations in which they were pursuing their research on "traditional" societies, only in reverse: Bruner was overtly a tour participant or guide while quietly developing an anthropological understanding, a process he unpacks only minimally (but see, for example, Bruner 2012, 154–55).

16. While there is but one mention of Clifford and Marcus's 1986 volume in Bruner's primary works (2005b, 6), he participated in numerous conference sessions bringing together the same group of individuals. See Di Giovine, this volume.

17. See Bruner (1999, 461–62; 2005b, 18–19) for discussion of additional, closely related concepts from non-anthropologists, including Mary Louise Pratt's (1992) "contact zones" and Homi Bhabha's (1994) "third space."

18. Bruner, personal communication.

Chapter Two

{Genealogies}
On the Emergence of Identity and Borderzones as Key Concepts

Kathleen M. Adams

The exploration of identity dynamics and touristic borderzones are two recurrent analytical targets in Edward Bruner's tourism research. This chapter argues that these foci are natural outgrowths of his earliest work in Native America and Indonesia. Going back to his days as a graduate student, Bruner's writings have revisited three analytical issues: mobility, inter-group encounters, and identity. Although Bruner has embraced varied ethnographic writing strategies and fieldwork settings over the course of his career, a consistent thrust of his work has entailed analyzing the creative interplay between these three conceptual realms. As I suggest in this chapter, Bruner's long-term engagement with these classic anthropological issues (migration/mobility, cultural encounters, and identity/ethnicity), in tandem with his "radically reflexive ethnography" (E. Bruner 2005b, 9), laid the foundation for his subsequent critical interventions in tourism anthropology.

Bruner's analytical contributions to our understanding of identity and borderzones are entwined with his ethnographic and methodological strategies. For tourism anthropologists conducting identity-oriented research in "small places" (Eriksen 1995), Bruner has modeled a pathway for adeptly moving from locally specific, vibrantly textured intercultural encounters to broader levels of analysis (see, e.g., K. Adams 2006; Leite 2017; Markula 1997). From his earliest writings to the present, Bruner's work has reminded us of the necessity of situating mobile persons in the context of broader historical, political, and economic undercurrents, thereby providing us all with the methodological scaffolding for analyzing intergroup relations, identity dynamics, and contests over meaning in and beyond touristic borderzones. This chapter examines Bruner's key contributions to our understanding of migration, identity, and meaning-making and underscores how this work has inspired scholarship in the anthropology of tourism.

ON THE TRAIL OF EARLY BRUNER: DISCOVERING NATIVE AMERICAN "BORDERZONES" AND CREATIVE IDENTITY-MAKING SPACES

Bruner's interest in cultural encounters, mobile people, and shifting identities dates back to his dissertation research on a North Dakota Mandan-Hidatsa community (E. Bruner 1954; Graburn and Leite, this volume). He succinctly described this early research in a 1989 biographical statement: "I studied Native Americans: why and how they changed, why some individuals changed more than others, who left the reservation for the city, and which aspects of culture were more persistent" (E. Bruner 1989a). His first dissertation-based publications were produced in an era in which acculturation studies were in vogue, so it is not surprising that his emphasis is on cultural transmission and change (e.g., 1956a; 1956b). Yet even in this early period, as with his later work on tourism, Bruner rejected approaches that simply describe differences in acculturation or "correlate them with a checklist of unweighted factors" (1956b, 605). Rather, from early on, Bruner interested himself in the *processes* whereby individuals and kin groups reaffirm and rethink their identities.

For instance, in his article "Primary Group Experience and the Process of Acculturation" (1956b), Bruner goes beyond simply identifying the presence of a "white model" (non-Native American relative) living in a Mandan-Hidatsa household as a significant factor in orienting a family toward the non-Native world. He unpacks fieldwork observations and interviews to showcase the processes whereby "parents consciously, deliberately and actively train their children to become like whites" (1956b, 619). His work takes us into the intimate world of "acculturated" family homes in which Native American parents push their children to speak only English and to socialize with other families that have turned their backs on the Native world. Likewise, in a move that foregrounds his long-term interest in the interrelations between performance and identity, he highlights the ceremonies and powwows avoided by acculturated families, noting that these public rituals "best symbolize . . . and make legitimate the traditional Indian values" (1956b, 607). Bruner gives his readers a suggestive sense of the identities at play in these ceremonies by drawing on the voice of an "acculturated Indian" man who tells him,

> I never go to any dances or pow-wows. These dances are fellows [*sic*] breeding grounds; they go there to drink, to raise hell, and they are not Christians. A lot of these fellows . . . go there to take some gal out . . . and get her drunk. That's all these dances are for, as far as the younger people are concerned. The older people go there to get these donations; something for nothing. . . . They take their kids and the kids don't learn much from that. They give things away and that money can be used to educate their children, buy provisions and things. It

seems like with an Indian they think they are in the spotlight; it makes them feel good. (1956b, 607)

In narrating what happens at Mandan-Hidatsa ceremonies and dances, Bruner's interlocutor is, in effect, crafting his own subject position as an "acculturated Indian" who stands apart from the Native world. Via early writings such as these, we can catch initial glimpses of some of the identity-oriented themes that were to blossom in Bruner's later work on tourism.

Central to this later work is a conception of performances and cultural displays as realms in which various identities (cultural, ethnic, national, and so on) are not only performed and narrativized, but strengthened, challenged, transformed, or rearticulated (see Bruner 1991; 2005a).[1] For instance, via Bruner's research on American tourists visiting New Salem, a reconstructed heritage site celebrating Abraham Lincoln (1993b; 1994), and in his study of Kenyans viewing ethnic displays at Bomas (2001), we see how the activities of domestic tourists strengthen sensibilities about "a patrimony that is already theirs" (2005b, 10). Yet Bruner also shows us that tourist sites do more than simply affirm identities; they are also spaces in which competing narratives about cultural identities and cultural patrimonies struggle for dominance—as he illustrates via his (2005a) analyses of New Salem, Masada, and Ghana's Cape Coast castles. In short, Bruner adeptly showcases the complexity of tourist spaces vis à vis identity dynamics.

I circle back to Bruner's theme of struggles over meaning and identity narratives in tourist spaces shortly, but first let us further examine his early work on Native American acculturation. From the outset, Bruner demonstrates the productivity of focusing on cultural borderlands in which individuals from different backgrounds come together and interact. For instance, his article on the varying degrees of acculturation in a Mandan-Hidatsa community identified interactions and intermarriages with whites as fertile realms for better understanding people's experiences of acculturation (1956b). Likewise, one of his initial Indonesia articles (1961) zeros in on the experiences of Batak migrants in the ethnically diverse city of Medan (Indonesia) in order to address theoretical debates concerning the impact of urbanization on ethnic identity and kinship relationships.

In his later tourism work, Bruner's focus on zones of intercultural interaction becomes more refined and reoriented toward spaces beyond the intimate realm of nuclear or extended family life: he targets what he terms "the touristic borderzone," which he defines as a creative space where "the natives have to break out of their normal routines to meet the tourists: to dance for them, to sell them souvenirs, or to display themselves and their cultures for the tourists' gaze and for sale" (1996b, 158; see also Simoni, this volume). Bruner's vision of touristic borderzones encompasses local-tourist interactions, habitual behaviors in these spaces of encounter, and even reflections about these encounters. Thus his analyses highlight an array of locales, ranging from sites such as Mayers Ranch, Kenya, where international audiences observe Maasai "tribal" dancing and share scones on the lawn with their British ex-colonial hosts (Bruner and Kirshenblatt-Gimblett 1994), to the moldy dungeons of historic Ghanaian slave castles, in which North American tourists on heritage pilgrimages to "Mother Africa" encounter local Fanti guides and performers (E. Bruner 1996a).

Bruner's pioneering work on touristic borderzones has inspired subsequent scholars to further explore these spaces of intercultural encounters, in an effort to better understand how stories and sensibilities about self and other comingle and take shape in the context of international tourism. For instance, Andrew Causey's (2003) work builds on Bruner's legacy by spotlighting marketplace interactions between international tourists and Batak carvers on Samosir Island (Sumatra, Indonesia). As Causey observes, this particular borderzone has the quality of a "utopic space" wherein both Bataks and Westerners have license to transgress their own cultural rules of polite behavior and playfully experiment with new identities and narratives. Causey's utopic tourist spaces have stage-like qualities: they occupy "a place in between reality and desire, a place that is both home and not-home, a place where things can be imaginatively re-created" (2003, 69). In short, certain touristic borderzones function as utopic spaces, rich in possibilities for emergent formations of self and other identities (see also Simoni, this volume).

MOBILITY AND MIGRATION IN BRUNER'S INDONESIA WORK

After Bruner completed his PhD, his post-dissertation research shifted to Indonesia, where he embarked on a project focused on what was to become a recurrent theme in his work: the interplay between mobility and identity. His first long-term ethnographic research on the island of Sumatra (1957–1958) centered on understanding urbanization and ethnic identity (see figure 2.1). Initially he spent seven months in a rural, highland Batak village researching kinship and change, but because a regional revolution provoked safety concerns, he relocated and spent the remaining five months studying among urban, "Westernized" Batak migrants in the coastal city of Medan (E. Bruner 1961, 508; 2005b, 8). Bruner's resulting publications productively questioned then-accepted truisms about the rise of secularism and individualism amongst rural-to-urban migrants. His analysis showed that Batak urban migrants maintain corporate lineage group affiliations irrespective of residence and that they returned to the village with their children "for important ceremonies and to spend vacations and holidays with their relatives" (E. Bruner 1961, 516). This prescient 1961 passage is perhaps the first passing mention of the dual dimensions of migrants' return travel to the homeland: they return both to vacation and to quench a thirst for family, heritage, and home. Still, Bruner's overarching objective in this early article is to identify variables that might explain why urban Batak migrants remained engaged in kinship organizations, contrary to the then-dominant theories of urban migrant identity transformation.

By the late 1990s, his research turned to a direct exploration of tourism-oriented identity dynamics. Bruner's interest in mobile subjects led him to chronicle the activities of Batak visitors at Jakarta's Beautiful Indonesia in Miniature Park (Taman Mini) in order to tease out some of the identity-oriented ironies of ethnic theme parks (E. Bruner 2005d). He noted, "ethnic theme parks are paradoxical because they display difference yet promote unity" (2005d, 212). Moreover, he tells us, the ethnic theme park recipe for promoting unity appears to vary from country to country, in accordance with a country's social demography. This chapter showcases Bruner's ability to productively travel between varied levels of analysis, moving from fine-tuned examinations of visitors' engagements with specific ethnic theme

Figure 2.1. Edward and Elaine Bruner with two Indonesian research assistants (an unidentified male and Minar Tobing) in a Toba Batak village, Sumatra, Indonesia, 1957.
Photo credit: Edward M. Bruner.

parks in Kenya, China, and Indonesia to broader cross-cultural comparisons. Ultimately, in this later writing, he sheds light on the varied ways in which national demography, cultural politics, and identity come into play in the touristic borderzones of cultural theme parks.

FINAL WORDS

Despite an array of topical interests ranging from kinship and migration, to texts, play, and stories, and finally to tourism; throughout his career, Bruner has rooted his inquires in richly textured, fine-grained ethnography. His evocative descriptions, such as his account of impatient, hungry tourists anxious to abandon a sweltering Balinese temple festival—an anthropologist's nirvana—in pursuit of lunch, conjure up empathy for all involved. But his vignettes also serve up important insights into the disparity between anthropological and touristic "ways of seeing" (E. Bruner 2005b, 7). As underscored at the outset of this chapter, Bruner does far more than simply offer evocative ethnographic descriptions and vignettes[2] about locally specific encounters in small places. Rather, his localized vignettes are breeding grounds for broader insights into the nature of human experience and meaning-making.

In following the path of Bruner's academic journey, we find that a staple in his scholarly suitcase has been the creative interplay between mobility, intergroup encounters, and identity. His attentiveness to these analytical targets in diverse settings and at varied analytical levels has ultimately fueled two of his critical interventions in tourism anthropology: an understanding of tourism settings as borderzones and a richer appreciation of identity dynamics at play in societies on the move. These critical interventions promise to pave the way for more nuanced, finer-grained understandings of topics of interest to a new generation of anthropologists engaged in critical tourism studies, including inequality, social justice, environmental justice, and tourism ethics.

NOTES

1. Bruner's earliest work on acculturation also reflects a vision of group identities as dynamic and fluid at a time when most scholars were still embracing more primordial understandings of group identities as fixed inheritances from the past. It was Bruner's approach to ethnicity as dynamic that informed my own dissertation field research on tourism, ethnicity, and art in 1984–1985.

2. I am indebted to Jean-Paul Dumont's insightful discussion of the expression *vignette*. As he observes, a vignette is not only a story but a "little vine" that twists, turns, and climbs to higher places (Dumont 1992).

Chapter Three

{Influence}
"So in Effect I Was Studying Myself"
Knowing (Our) Tourist Stories

Julia Harrison

In 1994, as a newly hired anthropology professor, I fretted over the viability of conducting research with domestic tourists. All sage wisdom would suggest that those with my pretenure status should simply build on an established research agenda to ensure that the requisite number of grants could be obtained and publications realized in time for the all-important tenure application. Never mind launching into a field of research that in the mid-1990s had limited visibility, not to mention credibility in the field of anthropology. If tourists were found in anthropological field sites, generally they were ignored. If noticed, they were likely to be disparaged, dismissed, and trivialized; at best, they were considered to be polluting irritants (see Errington and Gewertz 1989).

Regardless of these considerations, I had wanted to study tourists since the late 1970s. I had marked my determination to do so by purchasing Valene Smith's new edited volume *Hosts and Guests* (1977). Symbolically to me, it was the first scholarly book I acquired that was not a university course textbook. I lamented at the time that tourism and tourists were not something I could have studied during my undergraduate years in anthropology. The discipline had a long way to go before such courses were included in the curriculum.[1] Some twenty years later, after a diversionary career in another field, I had completed my doctorate and secured a tenure-track academic position, yet I still had not started on such research.

Shortly after completing my undergraduate degree in 1974, I had set out with my backpack for nearly two years of global travel to find out what it really meant to be a tourist. For better or worse, throughout my journey I was plagued with the mind-set of a fledging anthropologist. I never stopped wondering why I and many other young Canadians, Americans, Australians,

New Zealanders, Germans, and Brits were preoccupied with stretching our limited budgets, suffering endless bouts of "Delhi belly," enduring cramped and often not very charming hostels overseen by sergeant-like landlords, frequently missing out on important family events at home, and seeing few of the local sites. Our adventure was really about time spent with our fellow travelers (see Vail 2004).

These memories never left me. Nor did the emotional and bodily responses I experienced: the tingling I felt in my spine when I first saw the Taj Mahal lit by a full moon; or the time warp into which I was sure I had fallen when I saw an Afghani camel caravan shadowed in the moonlight on the horizon; or the olfactory assault those same camels offered when more closely encountered; or my incredulity at the massive Buddha statues in the Bamiyan Valley; or the bone-deep nauseating weariness as we rattled along on the wooden benches of yet another overnight train in Southeast Asia; or the intense panic and fearfulness I felt to my core when confronted with the masses of humanity in the streets of Mexico City. Just why did I choose to endure these things? I was never quite sure, but the emotional architecture of these experiences imbued them with endless capacity to fuel the stories I later told about my travels.

Throwing caution to the wind in my new academic position in the mid-1990s, I began a project to study a group of Canadian tourists to figure out just what made their travel experiences meaningful to them. My confidence to proceed was spurred on by trends that had emerged within the discipline in the late 1980s and early 1990s. Edward Bruner's seminal thinking about the wisdom of opening up anthropology as he (and his contributors and co-editor) laid out in *Text, Play, and Story: The Construction and Reconstruction of Self and Society* (1984b), *The Anthropology of Experience* (1986c), and later in his reflections on the writing of an ethnographic text in his introduction to *Anthropology and Literature* (1993a), were important contributions to these new directions (see also Marcus and Fischer 1986; Clifford and Marcus 1986; Van Maanen 1988). In the same time frame, Bruner joined a small coterie of other anthropologists whose work marked tourism and the tourist experience as something worthy of anthropological study (Bruner and Kirshenblatt-Gimblett 1987; E. Bruner 1991; Crick 1989; Nash and Smith 1991).[2]

The intersections of these two threads in Bruner's work presaged what Julie Scott and Tom Selwyn (2010) would posit over twenty years later: tourists and tourism are good for anthropology to think through both long-standing and emergent issues and debates about theory, method, and practice in the interconnected and highly mobile late capitalist world. They noted that anthropologists approach tourism in relation to a wide range of contemporary anthropological interests, seeing it as a practice "embedded within broader social, political, and historical frameworks" (2010, 17; see also Abram,

Waldren, and MacLeod 1997, 2). Bruner's important work on this topic exemplified if not inspired its interpretivist and humanist leanings, as observed by Naomi Leite and Nelson Graburn (2009) in their review of the field at the end of the first decade of the twenty-first century. But it also met Scott and Selwyn's (2010, 17) imperative to engage with the "political economy of anthropological knowledge," with specific reference to the study of tourism.

Bruner's frankness, honesty, and openness to the mere idea of being a tourist as both an observed and a deeply personal experience, his intense self-reflexivity about his (sometimes failed) role and his direct interventions in the experiences and responses of the tourists he went on to study, his insistence that narrative and story are central to human experience and expression, and his staunch belief that it is impossible to banish the persona and emotions of the ethnographer to "the edges and the margins" in fieldwork and the writing of ethnographic texts (E. Bruner 1993a, 2) were vital contributions to the move to finally take tourists and tourism seriously in anthropology. It is a fundamentally appropriate move if, as Jeremy MacClancy (2002a, 4) states, "anthropology is about taking people seriously."

In this chapter I reflect on how Bruner's boldly self-reflexive positioning cohered both conceptually and practically with my own research with tourists. The centrality of his distinction between "the trip as lived, as experienced, and as told"—or, in other words, what shapes the post-tour "story . . . that makes sense of it all"—was vital to my research (E. Bruner 2005b, 19–20). Woven into my narrative here are reflections on the role of emotions and affect: both those experienced by an anthropologist in the course of what some might call fieldwork and others would call a leisure activity, and those that shape the memories of tourists, informing fundamentally the stories they tell about their travels. I highlight the ongoing resonance of these themes in selected current writings on tourism by twenty-first-century anthropologists.

PLAYING THE TOURIST/BEING AN ANTHROPOLOGIST

At various points in his writings, Bruner is refreshingly honest enough to admit to being a "mere tourist." He observes and comments on the simple fact that anthropologists can be tourists. They can even like being tourists. And they need not apologize for doing so. He describes how in the 1970s he and his family went to Bali on a few vacations, taking time off from his fieldwork in Java: "We stayed at tourist hotels or beach cottages . . . and from what I recollect, we behaved in ways essentially similar to other tourists in Bali." He continues, "I thoroughly enjoyed these Balinese family vacations" (1995, 231).

In another touristic moment, in 1983, following a conference in Baton Rouge, Bruner and many of his colleagues attended Mardi Gras. He notes, "I was pleased to see a group of serious academics wearing sets of colored beads and outlandish T-shirts . . . jostling with the crowds and struggling to make sense of the . . . festivity that overwhelms an entire city" (E. Bruner 1984a, 1). Both the carnival and the papers at the conference, he later encouragingly observed, "were moved by the same spirit of openness and freedom" that was beginning to take hold in the discipline in the 1980s (1984a, 1). By the mid-1990s Bruner had the enhanced chutzpah to admit that once, in Ghana, in the midst of a festival, he found himself "dancing in the street." He writes, "I looked across the road at an old woman selling in the market, who was also dancing in front of her stand, smiling at me. Our eyes met. . . . It was a joyous moment" (1996a, 301). In these happy marriages of the anthropological self and the tourist self, Bruner boldly positioned the tourist as a real person, a curious and emotive creature who seeks pleasure, and not simply as a "polluting irritant" to the anthropologist (see Frohlick 2010). Engaging in a reflexive analysis of his cultural self, Bruner brashly juxtaposed the deprecated commercial, tacky, and superficial other—that is, tourists—and the stories they narrate for and about themselves, with the stories we tell ourselves about our own anthropological and tourist selves.

Bruner's willingness to engage with tourism and the idea of being a tourist emerged in the "reflexive, sometimes painful, always critical, and still evolving self-examination of the ethnographic enterprise . . . 'the experimental moment' in anthropology and the postmodern turn" of the late 1980s and early 1990s (E. Bruner 1993a, 1; see also E. Bruner 1995, 224; Marcus and Fischer 1986). He saw it as counterproductive to separate interpretations based on one's own "experiences in the field" from those based on "the experiences of others" (1993a, 9; see also Marcus and Fischer 1986; Clifford and Marcus 1986; Van Maanen 1988). Rather, a juxtaposition between the two can productively be the crux of one's field methodology, challenging the "model of the idealized anthropologist, as hero or victim, rather than as a frail all-too-human being" (E. Bruner 1993a, 9). Such ideas are now readily accepted; then, they were more radical, challenging fundamental assumptions about the objectivity of the position of the anthropologist (Collins and Gallinat 2010). But they helped nurture Bruner's anthropological study of tourism and tourists, for, to paraphrase the 1970s comic character Pogo,[3] we have been the tourist, and the tourist is us—even if we are anthropologists.

This acknowledgement is critical. It suggests the hesitation expressed by a young twenty-first-century anthropology student who, when he learned what an upcoming class would be about, exclaimed, "*Tourism?* But that's something you just do, and enjoy yourself, and not think about. It's not something you *study*" (MacClancy 2002b, 418; italics in original). Ironically, anthropologists have long acknowledged and written about leisure and hobbyist pursuits that have informed their professional research and written work. Dancers, actors, weavers, potters, motorcyclists, and recreational sports participants and fans were all willing to admit that their informal leisure participation played a role in what they studied and how they thought about it (see, for example, K. Adams 2006; Barba 1993; Dyck 2010; Kaeppler 1985; McCloskey 2002; Ness 1992; 1996; Skinner 2010). But it took considerable time for the leisure pursuit of being a tourist to be recognized in the same way, so virulent was the posture that the work of the anthropologist must not be conflated with the play of the tourist (E. Bruner 1991; Crick 1985; Ness 2002). Such recognition does not suggest that the anthropologist and the tourist are the same thing, but it does open the door for auto-ethnographic observations and for recognition of the role of memory and the multidimensional sensorial and affective character of the experience. It was also part of a shift in the discipline that made it acceptable, if not desirable, for anthropologists to focus their research energies on those at home, to accept the study of the Western self as not only credible but a vital element of the work of the discipline and part of "anthropology's poetic sensibility and graphic scrutiny" (Okely 1996, 5, 17).

The question that haunted those who might think about doing research on tourists and tourism was: How would an anthropologist do meaningful research with individuals among these hordes who are engaged in a fleetingly collective leisure experience, while at the same time struggle with the tensions (if not simple pleasures) of just being one of them? How does one productively think about ways of being a tourist-cum-anthropologist—or is it anthropologist-cum-tourist?

BEING AND GUIDING: THE BRUNERIAN EXPERIENCE

Bruner and others have readily acknowledged that the "key difficulty in studying tourists is methodological" (E. Bruner 1995, 225; Frohlick and Harrison 2008). Thus when Bruner and a few others started to do research on tourism and tourists in the 1980s and early 1990s, they quickly learned that their desired work demanded ingenuity, creativity, adaptability, and a certain degree of bravery. But by the late 1990s conversations had emerged in the corridors of the academy about the changing dynamics of anthropological research, and this dialogue expanded as the twentieth century rolled into the twenty-first. Anthropologists had begun to talk about "constructing the field" (Amit 1999) as research strategies adapted to global mobility and to the character of places where anthropologists did, or might do, research—complexities that had always been in place to some degree, but that anthropologists had either purposely ignored or more benignly overlooked throughout the twentieth century. Anthropologists also gradually became more comfortable with "studying up" (Nader 1972), acknowledging that if "people move, anthropologists [can] follow their movement to a new site, whether refugee camp, centre for migrant labour, or tourist destination" (MacClancy 2002a, 12).

Bruner's tourism research over thirty years examined many dimensions of the touristic experience: the ethnographer and the ethnographic study of this field; the experience of the tourist; the expert guide; the site, destination, or attraction; the locals; and the return home. His work, like that of few others, is notable for the manner in which he has consistently and deftly intertwined the ambiguities, tensions, paradoxes, and complexities that characterize all that happens in what he has called the tourist borderzone (E. Bruner 1996b; see also Simoni, this volume). He demonstrated that "these topics can best be studied by fieldwork and by the methods of ethnography" (E. Bruner 1999, 475). Or, as Scott and Selwyn echoed over a decade later, anthropology brings "a commitment to ethnography and ethnographic detail" to the study of tourism and the tourist experience (2010, 18). Some of Bruner's richest writings on tourism draw on his experiences of and reflections on acting as an expert guide. These practices influenced how I went on to think and write about two groups of Canadian tourists, even though our methodological strategies differed considerably.

Drawing on his knowledge of places such as Thailand, Burma, and Indonesia (including Bali and Sumatra)—destinations that, in the late twentieth century, came to be burdened by, yet productively embedded in, the dynamics of global tourism—Bruner was able to gain employment as an expert tour guide. He later expanded his work as a guide to select destinations in Africa. Many elite tour companies offer the services of noted scholars on their trips, frequently using retired academics, often anthropologists, and marketing these individuals as important value-added elements to their tourism products. But few anthropologists, Bruner notes, make these experiences into opportunities for useful ethnographic fieldwork or "incorporate [them] into their academic discourse"; still fewer ever write about them (1996b, 162; see also Dyck 2010; Collins and Gallinat 2010, 4).[4]

Bruner's career as a guide beginning in the early 1980s allowed him not only to travel with tourists, but also to follow up with them once they returned home (2005b, 26–27). He discovered that such a research strategy had both strengths and pitfalls. It can present the guide-cum-anthropologist with conflicting agendas. Are the tourists to be educated by remaining at a rare ritual event to witness its complete manifestation even if the timetable extends over several hours on a rather indefinite schedule? Or are they to be entertained by staying just long enough to be able to claim that they have "seen it" (2005b, 6–7), but not so long as to compromise arrangements for the next stop on the itinerary?

It is important to recognize that the imported expert guide is normally a position available only to those who have had the opportunity to accrue over years or even decades a significant depth of linguistic, cultural, and social capital as an expert about a specific place or tourist destination. It was such knowledge that got Bruner the job as tour guide, but it was his passion for sharing so much of his expertise that eventually got him fired from one contract (E. Bruner 1995; 2005b). That firing provided Bruner much fodder for reflection. He notes that the tour agency that had hired him was selling Bruner, the scholar, as much as it was selling the destination (1995, 226), so his professional self was integral to the stories of the tourists in his charge.

What does it mean to become a commodified "tourism product" (Bunten 2008)? Bruner (1996b) tells the story of guiding a trip in Bali during which he was able to conjure up anthropologists Hildred Geertz, Margaret Mead, and Gregory Bateson. Bruner's "attractivity" was enhanced when he and the tourists found themselves (thanks to the still-living Geertz) at the house where Mead and Bateson had lived decades earlier, and the tour group stopped to take pictures. It seems that Mead, Bateson, Geertz, and Bruner presented little ambiguity to the tourists. What they offered through their anthropological persona (ghostly or real) was their "scientific authority"—opening the door to the touristic pleasures of "the 'real' Bali" and elevating these visitors to something beyond "mere tourists" (1995, 226).

What did it mean for a tourist to actually have "seen it," as Bruner painfully remembered? Did it mean to have seen just enough, knowing that much of what was to be performed would be largely incomprehensible despite careful tutoring by a diligent and earnest expert guide? What did it mean, as Bruner would later muse, to "surrender" (1995, 237–38) to such self-knowledge—that is to be content with, if not desirous of, such limitations, to be happy, as one of those in his charge quipped, to simply see "the tourist dance"? Did having "seen" something imply that the tourist experience was primarily a visual one, at most a photo-op? Or was such "seeing" a richer shorthand intended to delimit an overly stimulated body and mind that had heard, smelled, touched, and tasted well beyond a comfort zone, prompting a bodily response informed by expectations based on age, national origin, and class regarding appropriate levels of noise, personal physical space, and privileged attention? What would the tourist remember? What emotions would such memories provoke? These are unsettling questions that continue to haunt any effort to understand the tourist experience.

COLLECTING MY (OWN) TOURIST STORIES

I lacked Bruner's depth of linguistic, cultural, and social capital about any "exotic" place that might be considered a marketable and desirable tourist destination. Bruner's research strategy to act as a scholarly tour guide was not available to me. It is frequently a strategy accessible only to those of a certain age and stature. It also can privilege access to a particular type of tourists. The social and cultural particularities and the type of touristic experience of those in Bruner's charge define them and thus their collective travel experiences as ones not regularly or historically studied by anthropologists (E. Bruner 1995, 227). They were clearly to be "studied up" (Nader 1972). Bruner readily admits that his only access to this elite demographic was as a tour guide, as like most academic researchers he did not have the financial resources to simply participate in such tours purportedly as just another tourist (even if he would have in reality been doing anthropological research) (1995, 225). Guiding such groups allowed him extended access to the experiences of these elite tourists. In his writing, Bruner is largely silent about the distinctiveness if not exclusivity of these experiences. He narrates the experiences but does not comment at length on the particularities of these stories as shaped by the privileged position of the tourists.

When I began research on tourists over twenty years ago, my first project was to gather the stories of individuals who had made a commitment to frequent international travel. My second, later, project studied those who were committed to a second-home domestic tourist experience. In both cases, these individuals had to be locally resident so I could visit them in person, often on a repeated basis. Both groups of tourists I worked with could be categorized as upper-middle or middle class, many were professionals, most had some postsecondary education, almost all made their permanent home in the Canadian province of Ontario, and all were passionately committed to the virtue of their chosen tourist experiences (see Harrison 2003; 2008; 2013). They shared some of the characteristics of many of those Bruner guided, but they were Canadian as opposed to American and were slightly lower in the socioeconomic hierarchies of North American society.

For my first research project I placed advertisements in both a national newspaper with a section for those seeking research subjects, and in a local information-sharing, readership-driven monthly travel magazine. I indicated that I wanted to talk to people who regularly traveled internationally for pleasure—whom I came to call "travel enthusiasts" (Harrison 2003). No single kind of travel experience characterized the trips of the people who responded and whom I ended up studying (see Harrison 2003; 2008). For my work with second-home tourists, I solicited contact with second-home owners through websites, newsletters, and information networks. Many friends and colleagues I knew fit the criteria of those I wanted to study, but I did not want to barter friendships to locate a group of domestic second-home tourists or cottagers for my study (Jaakson 1986). I focused on a region of the province of Ontario's "cottage country" that demographically mirrored the socioeconomic category of the travel enthusiasts I had previously studied. I was not going to do research where the wealthy owned cottages; nor was I going to venture to areas populated with more modest trailer or caravan parks that served as cottage communities for many working-class Ontarians (Harrison 2013). For both of my research projects, my subjects were self-selecting.

As an anthropologist, I sought to collect the richly textured stories of these two groups of tourists to gain some sense of where and how meaning was nurtured in their touristic experiences. On the basis of my own travel experiences, I knew one thing for certain for the first group: it is difficult to find anyone to share your travel stories with once you arrive home. This reality shaped my strategies for research with this group. Most friends and family are interested in only the most cursory description of your travels, signaling that they only want to hear selected high points. As an anthropologist interested in the tourist experience, I thought I could position myself as a keen, tireless, and enthusiastic audience for these stories upon their return home, and for their plans for future trips. Although I would not be talking to these people in the act of being a tourist myself, I would benefit from the reality that they would take the time to talk to me at length about their travels and plans. They also wanted to know of my own travel history and future travel plans. Thus I continually relived my own travel memories, both those I gathered in my mid-twenties and my recollections of ongoing forays into the tourist worlds of beaches, resorts, hiking trails, and cycle routes at home and abroad, as well as of the art galleries and museums of Europe. I was never allowed to leave these stories at home. They were my constant companions as I gathered the stories of travel enthusiasts. I relived these travel experiences with those I interviewed.

I visited the second-home domestic tourists at their cottages, the place where they acted out their touristic behaviours. But because I was an exceptional visitor to the cottage—I was not a family member or a close friend but was there on professional business—much of normal cottage life and activity was suspended while I was there. Most normal cottage activities—swimming, boating, making repairs, lounging on the dock—were temporarily postponed until after I left. Thus I heard about, rather than observed, what went on at the cottage. I was told stories about what went on there, frequently what had happened in the past because many wanted to tell me stories of their intensely emotional childhood memories of life at the cottage.

While doing this research my partner and I were based at a research center in central Ontario that had originally functioned as a cottage. This was my first exposure to cottage life. The cottage was built in the 1940s and retained much of the character and many of the elements of early cottages in the area. Thus it seamlessly blended in with the neighboring cottages and appeared to many on the lake to be one more of the same. My own extended family treated it as my—and thus their—summer cottage while I was resident and conducting my research. They regularly showed up for visits, and by the end of my first summer of research, I came to appreciate at far more than an intellectual level the idea of the cottage as a place where extended kin get to know each other; as a location set apart from the pace of the normal day-to-day world; and as a site where hours can be productively spent marveling at the complexity and beauty of nature—all things those I interviewed kept telling me it was. My sadness and regret when I had to leave on the last day of the summer were as raw as those described by the cottagers I had interviewed that summer. I had lost much of my intellectual distance from the cottage experience, but I had gained some enormously valuable affective insights.

COLLECTING BRUNERIAN STORIES

Studying tourism (and thus tourists) was a continuation of the trajectory of Burner's earlier work (see Graburn and Leite, this volume; Adams, chapter 2, this volume). He reflected that the shift coincided with his "personal intellectual development [and] changes in the discipline," and that it "perfectly incorporate[d] [his] interests in travel, movement, encounters, reflexivity, agency, contested narratives, and emergent culture" (2005b, 9–10). He had begun his tourism research at the moment in 1983 when he and his colleagues were "fighting for trinkets thrown from [Mardi Gras] floats" (1984a, 1). At the time he argued that anthropology was "opening up," becoming a field "that takes account of the spontaneity, improvisation, and innovation inherent in social practice, that sees culture as being in production, that is skeptical of fixed meanings and resolved endings and recognizes inconsistency and ambiguity of social life" (1984a, 13). This paradigm clearly informed his reflections during his other touristic moments, sowing the seeds for his constructivist approach to the tourist and the local's experiences. Bruner wrote that there was "a freer spirit on the loose in our discipline now, one that is more fun and more open, like a children's story or a mardi gras." This spirit, he continued, allowed anthropologists to "be more honest with ourselves and acknowledge the force of our own emotions, . . . to merge with our subject matter and open up not only to anthropology but to ourselves [and to] look forward to a more creative period" (1984a, 15). Such sanctioned liberation continues to echo in twenty-first-century anthropological study of tourism and the tourist.

To Bruner, anthropologists can tell their stories only from where they always are: "in the middle," a place that acknowledges its naivete and its lack of purity, positioned always between what has come before and what will come after. He asserts that "life consists of retellings" (E. Bruner 1986c, 12).[5] But what gets retold? Cultural stories (some might say master metanarratives) exist outside of particular narration (Bruner and Gorfain 1984, 56–58). But any individual narration captures in its expression a moment in a life, embedding one interpretation in a fluidity of meanings, forging a dialogic connection to a past, a present, and a future. Bruner has aptly observed that "stories may have endings, but stories are never over" (1986c, 17). He reminds us that narratives privilege the subjective experience of reality, often something different from the factual reality (E. Bruner 1986c; D. Picard 2012, 3). Such subjectivity is what makes a good story. But such good stories are not neutral. Tourist tales "privatize, embellish, and transform master narratives" (E. Bruner 2005b, 24).

Such stories also are more frequently about the experience of tourism than about any particular destination (Bruner 2005e). They do not necessarily reflect precisely what happens on the trip; on the contrary, as Bruner wrote in 1986, distilling Renato Rosaldo's (1986) reflections on the role of a "good story" among Ilongot hunters, "if one measure of the success of the hunt is the story that is subsequently told about it, then what happens on the hunt is partly determined by cultural notions of what makes a good story" (Bruner 1986c, 17). One need only substitute the word "trip" for "hunt" to capture Bruner's own insights about tourist stories. And, as he learned as a tour guide, he was part of what happened on the trip, and thus also contributed to shaping the "good stories" later told.

Bruner admitted ironically that through his lectures as an expert guide to the tourists in his charge, he influenced their reactions and thus their stories, the content of which he then documented in his fieldnotes. This realization prompted him to reflect, "in effect I was studying myself" (1995, 239; see also 1996b). The self here took multiple forms and played a role in multiple ways. Bruner was an "ordinary" American citizen (1995, 225), a tourist, and a professional anthropologist. In keeping with his American self, Bruner frankly admitted his comfort "with . . . affluent tourists because of the similarities of life experiences." Like the tourists, Bruner and his wife "were older college-educated professionals, and [they] too talked about their children" (1995, 227). He found himself negotiating the "ambiguous" and at times "uncomfortable" clash of his anthropologist-cum-guide-cum-tourist selves. It was difficult to separate these identities as he, the anthropologist, gaped like other tourists, "in awe . . . rushing from the bus to take photographs, enthralled by breathtaking scenery," but also "luxuriating in the hot showers . . . [and] complaining about the meals at the hotel in the evening" (1996b, 171).

These are not neutral comments. They reflect an approach to ethnographic research that embraces "explicit recognition of fieldwork as personal experience" rather than "sacrificing [that experience] to a false notion of scientific objectivity" (Okely 1996, 23). Such honesty requires anthropologists to "acknowledge the force of our own emotions" (E. Bruner 1984b, 15) in fieldwork, something Bruner (1984b; 1993a) first urged us to do decades ago. Much more recent writing on the subject takes up this cause again to "retrieve emotion from the methodological margins of fieldwork," where it seems to unrelentingly be positioned (Davies and Spencer 2010, 1). Emotions are integral to "lived experience," and making "analytical cuts between emotion and thought, or emotion, the senses, thought, and action" has been called a violent act (Davies and Spencer 2010, 25; see also Jackson 2010). To Mike Robinson, emotions cannot be ignored in the anthropological study of the tourist experience as they "are necessary, and in many ways inevitable, for engaging with the world" (2012, 22). As such, in accordance with Bruner, emotions cannot be excised from any attempt at understanding how we live in the world.

David Picard and Mike Robinson's edited volume *Emotion in Motion: Travel, Affect, and Transformation* (2012) privileges the place of emotions in thinking about tourism and the tourist experience. In his introductory chapter, Picard highlights Robinson's insight that emotions are "central, and yet often by-passed, in our understandings of tourism. . . . [T]hey are intimately bound to ways in which the world is imagined, mediated, and communicated and to the ways in which we perform, record, and recall our place in the world" (D. Picard, 2012, 15; see also Robinson 2012). At the most obvious level, recurrent tourists speak in their travel stories of destinations they loved, the passion they felt for people and places they visited, the sublimity of the fear they experienced, or even the strong dislike they felt for places to which they traveled (Robinson 2012, 31). These are not benign or generic responses. Frohlick adds another dimension to the emotional spectrum: female erotic desire—one of the powerful enticements that induced many of the Western women she spoke with to visit a Costa Rican beach community (2010, 52; see also Frohlick 2013). In my own work with two different groups of Canadians, nobody was dispassionate about the stories they wanted to share with me. The sheer willingness to put themselves forward to be part of my research was my first indication of their openness and desire to expose their deep connections to their cottages. To my surprise, tears flowed as some tourists shared their stories with me about things they had seen, even though I was a relative stranger to them (Harrison 2012).

James Davies and Dimitrina Spencer suggest that the generative space of emotion is "between things in relationship" (2010, 25). The tourists I interviewed impressed on me the importance of such spaces in what made their travels meaningful. Those who traveled internationally talked of how they valued "making connections," be they "the most tangential linkage [or] experiences of unquestioned intimacy" (Harrison 2003, 48; see also pp. 43–91). The domestic tourists who journeyed repeatedly to their cottages insisted that I understand that it was all "about family" there, a place that forged and reinforced tight bonds between "the nearest and the dearest" (Harrison 2013, 164; Pitkänen 2008, 182). Psychologists Linda Levine and David Pizarro argue that "emotion makes memory better" (2004, 537; see also E. Bruner 2005b), an insight that accounts for the powerful role emotions play in the treasured tourist stories I heard.

My own memories of my travel and cottage experiences fundamentally informed my research in both of my projects, even if my stints as an international traveler and as a cottager were separate from the experiences of those I interviewed (Harrison 2008). None of my interview subjects shared space with me as a cottager; nor did they come with me on my international travels. In some cases we did serendipitously travel to the same places, but our experiences of these places differed on many levels: the length of time we spent there, our accommodation choices, the size of our travel budgets, the company we kept while there, and our travel companions. Thus at an emotional, psychological, and visceral level my understandings of such experiences were entirely my own. This distinction, however, did not diminish the effect that my experiences had on how I came to interpret the travel stories I heard during my research. As Bruner had experienced, such personal emotive remembrances only strengthened my confidence in what I wrote about the stories I heard (see Harrison 2003; 2008; 2013).

Very specifically, my own emotional responses honed an understanding of what I have called respectively the "tourist aesthetic" (Harrison 2003, 92–138) or the "cottage aesthetic" (Harrison 2013, 41–42). I understood these concepts to encapsulate a range of emotional and aesthetic qualities, as follows:

> *the tourist aesthetic*—that which the tourist considers beautiful; that which gives sensuous, bodily pleasure and that which elicits positive emotional response; in something that the tourist sees from afar, as well as something in which she or he is immersed; things grand and things small; in the concrete and the abstract (2003, 95)

and

> *the cottage aesthetic*—the notion that the cottage is a fundamentally functional place in the shelter and facilities that it provides and also nurtures all that makes one feel good—embodying . . . the epitome of perfection, simplicity, and elegance . . . generated in an immersive, sensuous, and emotional bodily

> experience [that also] has resonances with a more distanced engagement. It is grounded in the simple, the mundane, the everyday lived dimensions of cottage life, along with the larger, awe-inspiring, and ultimately what are taken to be the sublime dimensions of the experience (2013, 41; see also Featherstone 1992, 268, 270; Leddy 1995).

I understood the resonance of the affect captured here, as I, too, had felt it. Put another way, I had both literally and metaphorically "been there." The treasured narratives told and retold by the American tourists Bruner guided and the Canadian tourists I came to know, while deeply meaningful, were far from neutral or monolithic. They were profoundly shaped, some would say disciplined, by a privilege informed by class, race, age, national origin, and imperialist legacies—by the larger metanarrative context, or what Salazar and Graburn (2014) would call "tourism imaginaries." Bruner and others would argue that the tourist is often blithely unreflective about such "narrative structures" (Leite 2014, 273; see also E. Bruner 2005b, 21; Salazar and Graburn 2014; D. Picard 2012, 3–4; Harrison 2003, 21–22). Leite recognizes such touristic imaginaries as "extraordinarily rich, collectively sourced resources for sense-making and world shaping," and yet, for tourists, stubbornly resistant to challenge (2014, 273).

If intense and visceral emotional experiences can be so potent as to lead tourists "to reflect [on] events of their life and rework their life stories," they are not something easily dismissed or disrupted (D. Picard 2012, 12). To understand and potentially shift this phenomenon, we as anthropologists need to be among the tourists, as Bruner charged. But we need to be fully among them in both our personal and professional selves. Only from such a dualistic yet entangled stance can we ever know the "poetic sensibility"—the affect, meaning, and potency of our collective stories.

NOTES

1. The latent nature of this desire fuelled my proposal during my first year of university teaching, twenty years later, to offer a course called the Anthropology of Tourism, even though there still were few resources available in the field.

2. The earliest anthropological work on tourism is Nuñez 1963. Nelson Graburn, a contributor to Smith's *Hosts and Guests* (1977), was one of the first anthropologists to seriously engage the subject of tourists and tourism with his work on tourist arts. He began to publish on this topic in the late 1960s; his key works include his edited volume *Ethnic and Tourist Arts: Cultural Expressions from the Fourth World* (1976) and his guest-edited special issue on the anthropology of tourism for *Annals of Tourism Research* (1983), which contributed to the foundation of the anthropological study of tourism. Theoretically and substantively this early work drew on the thinking

of well-established anthropologists such as Edmund Leach and Victor Turner, and those in sister disciplines like sociology (Cohen 1972; 1973; 1974; 1979a; 1979b; MacCannell 1976). Bruner and Barbara Kirshenblatt-Gimblett's article on East Africa (1987) likely had limited distribution. A revised version of that paper was published as "Maasai on the Lawn" in *Cultural Anthropology*, where it became very influential (Bruner and Kirshenblatt-Gimblett 1994).

3. Pogo was both the title and one of the central characters in a long running mid-twentieth-century American syndicated comic strip that frequently took on social and political issues of the day. Pogo's famous pronouncement, "We have met the enemy, and he is us," was made as he was depicted collecting scattered garbage on Earth Day, April 22, 1970 (Kelly 1972).

4. In recent years more of those who have acted as professional guides have begun to write about their experiences (see, for example, Guldin 1989; Bunten 2008; D. Picard 2012; Feldman 2016), but none documents the guiding experience of a noted scholar brought to a destination by an external tour company.

5. Bruner (1986c) notes his debt to Wilhelm Dilthey here.

Part III

EXPLORATIONS: CONCEPTS AND DEBATES

Chapter Four

{Authenticity} "Whatever We Weave Is Authentic"

Coproducing Authenticity in Guatemalan Tourism Textile Markets

Walter E. Little

For weeks, Ixkem, a Maya weaver from San Antonio Aguas Calientes, Guatemala, fretted over an enormous tablecloth she was weaving on her backstrap loom. Tablecloths are typically woven on treadle foot looms, imported by Spaniards in the sixteenth century and still used in Guatemalan towns. Ixkem's American client knew that in Guatemala treadle-loomed textiles are called machine made, while backstrap-loomed textiles are handmade. He wanted to purchase "an authentic, handmade tablecloth," big enough for his twelve-seat dining table, not something that was "machine made." He wanted her to weave a tablecloth with "traditional Maya designs, not flowers, angels, and parrots," which are typically found on *p'ot* (blouses) and *su't* (shawls and rectangular utility cloths) that women in Ixkem's town weave for themselves. Such commoditizing of Maya culture by Mayas and by tourists is common (Castañeda 1996; Little 2004; Medina 2003).

With these stylistic parameters, Ixkem set out to create a tablecloth according to her client's aesthetic tastes and notions of Maya culture. The challenge, Ixkem explained, was to weave something not typically woven by backstrap weavers that would be accepted by a foreigner with strong ideas about what was really Maya. I asked her if she thought the tablecloth was "really Maya" (*ri kan qitzij Maya'*). "Yes, of course," she replied without hesitation. "The difficulty (*ri k'ayewal*) is divining (*xtinwïl*) if he will think it is."

I begin with this example to explore authenticity through Edward Bruner's lens. He contends that "the local production of meaning is the format of dialogic interaction between the interpreter and small groups of tourists," a position he arrived at from ethnographic research at the historic site of New Salem, Illinois (E. Bruner 1994, 410). Bruner's conceptualization highlights the tensions that exist in cross-cultural coproductions of authenticity. This way of understanding authenticity emphasizes a dynamic process that does not rely exclusively on experts, be they anthropologists, museum curators, or others. It is grounded in the constructivist intellectual tradition—"culture is emergent, always alive and in process" (1994, 407). When I was a graduate student of Bruner's in the mid-1990s, this approach to cultural analysis helped me understand how to approach Mayas' conceptualizations of authenticity as they interacted with tourists, textile collectors, and other Maya weavers. Bruner had a playful attitude about authenticity. I recall him explaining that everything was authentic as he took apart arguments by experts who saw objects as intrinsically authentic or inauthentic. Ironically I left those seminars less interested in authenticity and more concerned with Mayas' political-economic practices as they related to instrumental uses of identity (Little 2004). It is with some trepidation, then, that I pursue this topic, given that Bruner stated in a 2005 interview, "the issue of authenticity has been overdone in the tourism literature. Authenticity is a red herring" (Lynn 2005).

My case study here differs from Bruner's studies (and most touristic studies of authenticity) because I focus on Maya hosts' opinions, rather than pursuing tourists' perspectives. I frame Maya weaver-vendors' and foreign tourists' cross-cultural interactions before analyzing how authenticity is coproduced in those interactions. In the discussion below, I revisit Bruner's uses of authenticity in order to reassess the significance of coproduction.

As noted by Eric Gable and Richard Handler (2005, 125), there has been a steady flow of authenticity and tourism research ever since Bruner began to give tourists serious ethnographic attention. My discussion of authenticity comes from the perspectives of those who work within tourism spaces, rather than those who consume tourism spaces. My study builds on and complements Bruner's critique of authenticity where he uses ethnographic evidence to challenge the modern tourist's quest to find the authentic (1994; 2005b; 2005e). The focus here is not on tourists searching for their own or other's authentic selves. Rather, I refocus Bruner's lens of authenticity critique on people who perform for and sell items they create to tourists.

I illustrate how conceptualizations of authenticity engender discussion and inspire economic strategies among Mayas, while they are not adopting Western meanings of authenticity for themselves. My discussion parallels that of Benjamin Feinberg (2006), who bases authenticity on Mazatec Oaxacans' perspectives. Because his subjects (Feinberg 2003)—Oaxacans and tourists—are embedded in different kinds of tourism practices and power relations, authenticity in this chapter differs. Nonetheless, just as Feinberg builds on Bruner's theory to illustrate how authenticity works on the ground, I pursue a similar strategy in Maya touristic spaces.

In the subsequent section, I summarize the anthropological debates about the concept of authenticity, then provide detailed ethnographic examples of Maya weavers' and vendors' interactions with tourists to discuss how authenticity does and does not play out. In that section, I provide a running analysis, drawing on Bruner's theorization and critique of authenticity in tourism settings. Before concluding the chapter, I analyze the cross-cultural Maya-tourist exchanges via Bruner and Dean MacCannell's debates about authenticity. This chapter extends and echoes Bruner's assessment of authenticity as a red herring by showing how the Maya weavers with whom I work also dismiss authenticity as a social scientist and tourist problem.

AUTHENTICITY DEBATES

Bruner's contributions to anthropological understandings of authenticity are located within theoretical debates that have preoccupied ethnographers for the last forty years. The literature on authenticity, even limited to tourism scholarship, is vast. Naomi Leite and Nelson Graburn's (2009, 43–44) concise, clearly written anthropological review is a good place to begin, but numerous tourism scholars have written comprehensive critical reviews on the subject (Cohen 2007; Lacy and Douglass 2002; Olsen 2002; Resinger and Steiner 2006; Shepherd 2003; Theodossopoulos 2013; Wang 1999). These reviews emphasize MacCannell's (1973) contributions, especially noting his concept of staged authenticity. To make clear Bruner's position on authenticity, it is important to outline MacCannell's conceptualization, before explaining how my analysis aligns with Bruner's (1994; 2005b; 2005e).

MacCannell's concept of staged authenticity (1973) melds structuralism with symbolic interactionism that is not far from Victor Turner's (1969) liminal ritual phase, except that in tourism this is in a public, not private space. Bruner, I contend, reacts to this structuralist orientation in order to find an alternative vocabulary to explain the experiential dynamics of touristic spaces. MacCannell's structuralist orientation is evident when he delineates six stages of a front-back continuum (1973, 598), basically refining Erving Goffman's (1959) front stage–back stage binary, which separates performers from their audience. The audience is excluded from the back stage, where "illusions and impressions are openly constructed" (Goffman 1959, 112). MacCannell argues that staged authenticity is a place between Goffman's front stage and back stage, noting that even Goffman admits that under some situations "the back-front division no longer allows one to make facile distinctions between mere acts and authentic expressions of true characteristics" (MacCannell 1976, 95–96). Later, MacCannell (2008) argues that the concept of staged authenticity goes beyond Goffman's front-back stage to capture a complex sociological dynamic that does not require analytical adherence to the idea of a real, more authentic back region. Instead, it refers to the creation of a touristic experience marked by the "pretentious revelation of back-region secrets" and a "constructed sense that [visitors] 'got in with the natives' or experienced local life as the natives experience it" (MacCannell 2008, 337).

In Bruner's (2005e) and MacCannell's (2008) authenticity debate, Bruner pushes us to reconsider that MacCannell's tripart division—front/stage/back—is arbitrary and constructed. This division, I argue, is multiply constituted, is produced by those who participate in tourism spaces, and depends on the context and the extent of mutually shared knowledge and culture. Being in the same place, regardless of conceptual frameworks of staged authenticity, is not enough for the space to be mutually intelligible by all parties. The same applies to material goods, such as Ixkem's tablecloth, irrespective of the commercial transaction that binds the weaver-vendor to the tourist-consumer.

Bruner's (1994; 2005b) position is that authenticity is typically built around arguments that essentialize and limit the broader, changing contexts in which tourists and those who host, entertain, and serve them participate in tourism spaces. For this reason, I take issue with MacCannell's (1973, 595) claim that "Touristic openings into society's back regions are ubiquitous." This presupposes that tourists and those who host them, entertain them, and sell material culture to them likewise conceive of a world partitioned into front and back, and that those spaces can be mediated through forms of staged authenticity. Given what I know of Mayas in Guatemala, I am suspicious of this universalizing of experience and meaning. This point is reinforced by Julie Lacy and William Douglass (2002, 6), who contend, "Both the native and the anthropologist, then, may perpetuate a static view of *the* authentic culture" (italics in original). Thus the gains made through the delineation and promotion of specific Basque ethnic expressions may compromise the ability to participate in wider political and economic spaces (Lacy and Douglass 2002, 8). This leads precisely to Bruner's (1994, 401) point, referring to four of authenticity's meanings—verisimilitude, genuineness, originality, and authority—that "the problem with the term *authenticity*, in the literature and in fieldwork, is that one never knows except by analysis of the context which meaning is salient in any given instance." It is through social coproduction that authenticity may or may not prove salient or important, as my examples below illustrate.

UNDERSTANDING AUTHENTICITY VIA MAYA WEAVERS AND VENDORS

The tourism literature emphasizes that tourists travel in quest of authentic selves. In Guatemala, foreign tourists tend to not question the authenticity of their selves. If they have these preoccupations or identity anxieties, they do not demonstrate them in Guatemala. And Mayas simply do not question their own authenticity. Bruner critiques Gable and Handler's (1996; 2005) position on this, arguing, "I do not believe that most tourists feel they have inauthentic selves and that through tourism they seek the really real, an authentic experience out there in the past, the old country, or the primitive. My data just do not support it" (2005e, 202). Elsewhere he observes that "MacCannell and Handler say that tourists are looking for authenticity, but it may be these contemporary intellectuals who are the ones looking for authenticity, and who have projected onto the tourists their own view of themselves" (1994, 408).

My research conclusions, here and elsewhere (Little 2004), concur with Bruner's. Like Bruner (2005b; 2005e), I did not find tourists struggling with self-identity authenticity. By contrast, some tourists worried about the characteristics of textiles that did or did not make them authentic, especially if they wanted to buy them. Hence, it is possible to explore the coproduction of authenticity via Maya textiles. Over two extended ethnographic examples with Maya weavers, I examine the concept and use of authenticity. In the following examples, I identify a fundamental challenge to vendors and an ideological conflict between vendors and their foreign customers with respect to authenticity—whether it is a process or an intrinsic quality of a thing.

Ethnographic Example 1: The Tablecloth

Ixkem is a Kaqchikel Maya weaver and textile vendor from San Antonio Aguas Calientes, a town located about a twenty-minute drive from Antigua, Guatemala. The women there are well-known and well-respected for their weaving skills. Tourism guidebooks have featured them for decades, and they have been the subjects of multiple scholarly studies (Annis 1987). The proximity of Antigua, Guatemala's most visited tourism destination, and the popularity of their own town on the ethnic and historical tourism circuits mean San Antonio residents have had regular contact with foreign tourists since the early 1900s (Little 2004).

San Antonio weavers, mostly women but some men, are widely known for their backstrap loom weaving skills. Ixkem is an exceptionally skilled weaver with a talent for innovation. She imagines new designs, recombining various patterns and colors that cut across stylistic generations of textiles, conceiving of new textiles that would have market value. She can weave the town's most conservative traditional styles, but her weaving interests rest more squarely in the tourism marketplace where there is more flexibility for her to experiment. Although, as Carol Hendrickson (1995) notes, most studies of Mesoamerican weaving focus on traditional textiles, weavers test out new combinations of colors and designs that articulate with local conventions. Some are embraced and copied, while others are one-off failures

Figure 4.1. Kaqchikel Maya weaver from San Antonio Aguas Calientes demonstrating backstrap weaving to students.
Photo credit: Walter E. Little.

that end up stored away and never worn again, or they are refashioned into something that can be sold to tourists—change purses, hotplate pads, decorations for vests, or other items.

Ixkem keeps her designs and new product ideas to herself, as do other weavers known for innovating, and she rarely lets her competitors see, much less handle, her products. The problem for her and others is how to connect with tourists who frequently complain about the sameness of the goods sold in market stalls. Thus, when new products are displayed, those that sell are copied. Copying is *de rigueur* among weaver-vendors, especially when they know a product is popular with tourists. This practice can be found in many other places in Oaxaca (Brulotte 2012; W. Wood 2008), Ecuador (Colloredo-Mansfeld, Antrosio, and Jones 2011), Indonesia (Causey 1999), Yucatán (Castañeda 2004a), and probably elsewhere. Even the moderately competent weaver can read most textiles and reproduce them. "You have to be careful," Ixkem tells me. "As soon as your weaving is copied, it is everywhere in the market. You have nothing special to sell and your weaving won't stand out. You lose your advantage."

The tablecloth was a particularly challenging project for Ixkem. It was longer and wider than typical backstrap loom weavings. She had to weave several long pieces, *paraj* (roughly twenty-four-inch-wide bolts of fabric), making sure the width of each *paraj* was consistent. Added to these physical and technological challenges, the design elements in the long *paraj* needed to line up. She expected that the designs she was incorporating—the *kumatzin* (a zigzag pattern), the *rupam läq (*a trapezoidal shaped pattern), and others—would match perfectly. She had to negotiate the problem of how to satisfy the buyer's notions of authenticity. She knew that the lack of imperfections would make it look like a factory woven textile; at the same time, too many imperfections would make the tablecloth look amateurish.

When the tablecloth was finished, Ixkem invited me to her home's inner patio, the only area of her house where she had room to unfold it. Everything about the tablecloth was impressive: the incorporation of the community's oldest textile designs, the mix of brilliant colors of *kumatzin* and *rupam läq* on a finely loomed base of off-white cotton mixed with threads of hand-spun brown, and the technological feat of weaving something so big on a backstrap loom. It was stunning and it took me a moment to express how beautiful it was. To fill the silent pause, she joked, "Well, this is something that no one will copy!" As we talked, she mentioned the American client who was purchasing it, growing more anxious. "Do you think he'll like it?" she asked me. "Will he think it is traditional or too different? Will he pay for what it is worth or what the threads costs and my time? Would he think it is authentic, something Maya?" I could not answer her questions and suggested that it would be better to see if he honored the economic agreement and that, irrespective of how traditional or authentically Maya he thought was, he would think it was a beautiful, skillfully woven tablecloth.

When the American client came to purchase it, Ixkem insisted that I be there to interpret. As Richard Adams (1970) recognized in Guatemala, cultural broker roles, such as the one into which I was thrust, are positions of power notably held by non-Maya Guatemalans who gained economic and political advantages mediating between Mayas and non-Mayas. Similarly, Kathleen Adams (1984) argues that travel agents play significant roles in the ways tourists understand the cultural ways of the people and places they visit. Bruner (1995; 2005b) too, reflects on his role as a culture broker when he was working as an anthropologist-turned-tour guide.

My role as a cultural broker parallels that of Hildred Geertz when she acted as a broker to get Bruner's tour group to purchase from a Balinese artist (E. Bruner 1996b). Geertz positions herself as an authority on Balinese culture, and Bruner, also an authority, sees himself as being imbricated in the tour agency's "marketing ploy to have a built-in authenticator"; similar to the

Balinese, they are "tourist objects" themselves (E. Bruner 1996, 170). Ixkem put me in this role because she wanted to use me to leverage her position and to help her better read her client. She and I spoke Kaqchikel Maya during the transaction and he and I spoke English and she and he spoke Spanish. He had few questions for me, other than to ask why I was there and who I was. He was happy to learn that I was an anthropologist and that Ixkem's work was considered worthy of anthropological study. "You probably have learned what are really traditional textiles, what is authentic," he told me. I responded that "traditional patterns or those that have long histories of use are not the same as authentic." He waved off my comment, not interested in what I thought was authentic, and turned his attention to the tablecloth.

Although he acknowledged that the tablecloth was outstanding, exquisitely woven, and beautiful, he was not "sure it was traditional enough. It just doesn't feel authentic." He said he was still interested in it, but they would have to discuss the price. The price was going to have to go down, since, as he put it, "it just wasn't the traditional tablecloth I wanted." Ixkem was furious. She said to me in Kaqchikel Maya, "What does he mean, not traditional, not authentic? This is very disrespectful and we had an agreement." More composed, holding the tablecloth, she turned to her client. "Look at this figure, the *kumatzin* (zigzag pattern). It is one of the oldest and most important designs we weave. It's from before the Spaniards came. And so is the *rupam läq* (trapezoidal pattern) and the *ch'umil* (star). How can my tablecloth not be traditional?"

"Well, there is nothing like it," he answered. "No one has a tablecloth like it. You just don't see tablecloths like this. It's not authentic. Unique, yes. Artistic even. But not authentically Mayan."

"I wove what you ordered," she responded and flatly added, "it is traditional and it can only be authentic." She called one of her daughters to bring several *po't* (hand-loomed blouses) and *su't* (shawls). She showed him the same design elements that were in the tablecloth and she explained the process of weaving to him, arguing that weaving on a backstrap loom in the time-honored techniques of her people meant that "whatever I weave is authentic."

"No, just because weaving is authentic, doesn't mean the tablecloth is authentic or traditional," he countered. "Lower the price."

She called to her daughter to bring her loom, already with a new project in process, and explained weaving fundamentals, but he was not interested. "I've seen weaving," he said. To him weaving processes may be authentic and a traditional practice, but that did not necessarily lead to traditional or authentic textiles. By contrast, tourists who are interested in Maya culture and history do get wrapped up in what is authentic, especially in the case of material culture like textiles. Many worry about being cheated, but more often their anxieties are about whether real life is being performed (see Little 2003; 2004; 2008) and if the things they are buying are authentic. As Bruner notes, authenticity "is a matter of power, of who has the right to authenticate" (1994, 400). This, however, is a complicated process with respect to Maya textiles.

Many Maya women working in the tourism sectors as vendors, as domestic servants, and for tourism companies see how they are represented by the tourism industry and are aware of power differentials. Once, when some Maya vendor women and I were looking at tourism brochures with photos of Maya women weaving, going to the market, and holding babies, one woman said to me, "These are just *mo'so'i'* (foreigners; non-Mayas), trying to make some money. If these pictures attract people to Guatemala, then maybe they will buy something from me." Another commented, "Photos like these are good, just as long as we don't look dirty or dangerous or crazy." When authenticity came up in conversation, it was in relation to sales of textiles and, sometimes, to cultural performances. In the latter, the tourists made comments among themselves and to their guide, but they did not mention to a performer that a performance was inauthentic. However, because the textile economic exchange was direct and most vendors were Mayas who wove, tourists did have questions related to the authenticity of the item during sales negotiations. The interaction between Ixkem and her American client illustrates that the authenticating process does not necessarily rest in the hands of either party. Neither person has hegemony, and impasses over authenticity can result in a satisfactory economic exchange, as eventually occurred between Ixkem and her client.

Frustrated by the stalemate over tradition and authenticity, Ixkem abandoned her argument that backstrap weaving makes the products traditional and authentic. Since he had provided her with a cash advance for thread, something she insisted upon, she produced receipts and said that he would have to cover the remaining balance. She took a ledger in which she recorded the times that she worked on the tablecloth. Having experience selling to foreigners, she knew that documenting her work would make a more convincing argument.

In the end, he agreed to pay her asking price. By his own admission, the tablecloth was too nice to pass up and, as he explained, "I know that I wouldn't have gotten the money back that I had already paid her. It's a beautiful tablecloth, really distinctive. It will look nice at home, but it's not authentic."

When he left, I asked Ixkem in jest when she was starting her next tablecloth. "Never! What a waste of time and way too much work," she replied. She said that she would go back to smaller, less expensive and time-consuming projects. To rib her a bit more, I asked, "Are you worried about tourists thinking these aren't traditional or authentic?" To which she responded, "Tourists have their own ideas about that. I'm a weaver from San Antonio. We're famous. Everyone knows us. What we weave is traditional, is authentic—whatever we weave." (*Pa ruwi' ri kem, ri turistas nkinojij achike chi nkajo'. Rïn in ajkem pa San Antonio. Oj kemonela' oj famosas. Konojel ketaman qawäch. Achike nqakem röj k'a ri' ri kan qitzij kem—xab'achike nqakem.*) She paused briefly and added, "What I should have done is showed the tablecloth to the other vendors and they could lose time weaving trying to copy it." (*Achike xinb'anataj xink'üt ri mantel chi re jalojoj chik k'ayinela', k'a ri xketikir xkisäch kitiempo b'anon jun juman.*)

Ixkem was not interested in discussing authenticity. She used the term *auténtico* (authentic) when quoting tourists and *ri kan qitzij wachinäq* (the real thing) when conversing in Kaqchikel Maya with me. "That was something the gringos worry about," she remarked. "I only have to think about it when I sell to them. Sometimes they care; sometimes they don't. One customer says it's not authentic; another says it is. They don't know what is authentic and I don't care."

In marketplaces around Antigua and elsewhere in Guatemala tourists question items' authenticity. This authenticity discussion is insulting to the vendors and weavers. "How can our weavings (*kem*) not be 'real weavings' (*ri kan qitzij kem*)?" they would ask me. "Weaving is what we do; who we are. That makes it real (*ri kan qitzij*)." As annoying and time-wasting as these conversations are, they do lead to sales and are reliable ways to gather information about tourists, even aiding in the development of new products that would be accepted as authentic by tourists (Little 2004).

Commissioned work is always risky, so even though Ixkem was skilled at such negotiations, she tended to avoid them. Vendors with less experience did not always conclude their business deals with foreigners. In one case, another weaver wove a larger tablecloth than Ixkem, incorporating conventional designs from her hometown of Chajul. When she finished her tablecloth and presented it to her client, the woman from Germany refused to pay her, claiming she had wanted "an authentic tablecloth" and that this was "definitely not authentic."

Authenticity, a concept imposed on the weavers, is negotiated in much the same ways that Bruner describes throughout his research (2005a). This construction by tourists, tour guides, and tour agencies can lead to impasses about what is authentic, as in Ixkem's tablecloth situation. Big items and large orders of smaller, easy-to-weave items can be too economically risky for vendors, whose products may be considered inauthentic by their clients. With simple items made with inexpensive thread, vendors respond more quickly to tourists' expectations. Many vendors were surprised that few tourists were interested in buying hand-loomed Maya clothing. More tourists purchased used clothing that was refashioned into other goods: pillow cases, purses, and other items. Some of the tourists I spoke to, who purchased these refashioned goods, said they were happy to have something "authentically Mayan" or "authentically Guatemalan" or "authentically Indian." Vendors cared about the sale and whether the tourists were happy with their purchase, irrespective of authenticity.

Ethnographic Example 2: Performing Weaving

Kaqchikel Maya vendor-weavers selling textiles in Antigua weave for tourists. Some invite tourists to their homes, and others strike deals with the owners and managers of hotels, restaurants, and Spanish-language schools to let them weave in lobbies and atriums (Little 2004). From Oaxaca, Mexico (W. Wood 2008) to Taquile, Peru (Zorn 2004), such public weaving for tourists is common in Latin America. These are clearly staged performances, but to conceive of them as forms of staged authenticity that, as MacCannell argues, tourists seek out as authentic experiences (1973; see also Chhabra, Healy, and Sills 2003; Desmond 2001; Harkin 1995), would be to miss what is happening. Embedded in Bruner's (2005a) work is a critique of authenticity

Figure 4.2. Students trying on clothing in a Kaqchikel Maya home.
Photo credit: Walter E. Little.

that I further develop: tourists do not go looking for authentic Maya weavings and weaving performances; this is not what authenticity is about or what is most important about these encounters.

Ixch'umil sets up her loom in a Spanish-language school, working most weeks Monday through Friday. Many women from her town of San Antonio Aguas Calientes also weave and sell in Spanish schools, hotel lobbies, and other places in Antigua. She typically begins her performance when the students, all foreigners, take their mid-morning coffee break from the language classes. She arrives as they take their first break, greeting them in Spanish and teasing them to learn Kaqchikel Maya. The student-tourists watch her unroll her loom, attach one end to a hook on the wall, and strap the other end around her waist. Many of the students stand silent, especially the new arrivals, who just observe her weave, while more advanced students joke with her, asking questions about weaving and family life in Guatemala. She is patient with them, gently encouraging the nervous language learners.

While most students are content to watch and chat with Ixch'umil, some try their hand at weaving, taking a turn for a few minutes, then commenting, "This is far more difficult than I thought," or "Five minutes and my back is killing me!" Few take up the opportunity to have weaving lessons with her. She explained to me that she does not like to take on many students. She is not protecting her back-stage self from the front-stage performances she gives at the school. "No," she tells me, "I don't give many lessons because I cannot make as much money as selling at the school." Indeed, when a student asks her for weaving lessons, she directs them to another member of her family. "Weaving is hard for foreigners," she continues. "Their bodies aren't used to it and most don't have the patience. Here, the students can play. They can try to weave and have a bit of fun."

Because Ixch'umil weaves each day the school is in session and students at this school must sign up for a minimum of five days of instruction, they see each other regularly. Students tend to take two or three weeks of Spanish classes, so they have many opportunities to converse with Ixch'umil and watch her weave. Students chat with Ixch'umil during their class breaks, taking brief moments to weave at lunchtime or when their class is over for the day. She and the students talk about their respective lives, and little relates to the topic of authenticity. However, authenticity is explicitly discussed as a conversational topic when students look at and consider buying Ixch'umil's textiles, as the following example illustrates.

Anne, a thirty-something female student from California, took Spanish lessons for two weeks. She watched Ixch'umil weave, chatted with her as her Spanish skills improved, tried to weave, and looked at the textiles. She responded like most students, who pick up the textiles, feeling and examining them with serious looks on their faces. Their questions tend to be: "Did you weave this?" "How long did it take to weave this?" Only when they are ready to buy do they ask, "How much is this?"

In the beginning, leafing through the textiles for sale and holding one up, Anne asked Ixch'umil, "Is this one traditional? I'd like something traditional."

"All of the weavings are traditional," Ixch'umil replied.

"But this one looks different from the others," Anne commented, adding, "It's really different from your *huipil*" (blouse, in Spanish).

"True" (*qitzij*), said Ixch'umil in Kaqchikel Maya. "There is a lot of tradition."

"Oh, okay." Anne said. When I overheard her talking with one of her classmates, she commented, "I don't think some of her textiles are traditional. Some just don't seem authentic. I'd like to buy something that's really Maya, but something Ixch'umil made."

A couple of days later, Anne asked Ixch'umil to show her which textiles she had woven. Ixch'umil gave her a table runner that is commonly sold by vendors from San Antonio Aguas Calientes. They contain various design elements, like the *kumatzin* (zigzag) described above, that are found in many Guatemalan Maya textiles, but the main designs on this piece were stylized quetzal birds. Other table runners had flowers, and the Christmas version had candles and mistletoe. Anne was disappointed in the textiles, saying, "This is something sold to tourists. I want something you would use."

"I do use these," Ixch'umil answered.

Anne continued, "It just doesn't seem," taking a long pause, "well, really Maya."

"Everything I sell is really Maya," said Ixch'umil. "We weave all the weavings. They're all real" (*Todos son verdaderos*). She showed Anne another piece, a *tzätz* (baby cap). The small rectangular textile would be sewn into a conical shape, adjusted to the size of the baby's head. That *tzätz* was a basic version, predominantly red in color with a ring of *ch'umil* (stars) around what would be the rim of the cap.

Anne seemed pleased with this option but asked to see another one. Ixch'umil gave her a double-sided *maracador*-style *tzätz*. This weaving style is inspired by brocade techniques that Maya weavers adapted to backstrap looms. Anne liked this piece. "This looks like something you wear," she told Ixch'umil. "What is it?"

Ixch'umil took the *tzätz* and showed Anne how to sew into a hat. Then it made sense to her. "Oh, I thought you made something for tourists, not something you'd use." She then asked about the price. It was almost fifty dollars. Her simple "oh" and downcast eyes revealed her disappointment and surprise about the price. "It's really expensive. How much is the red one?"

"That is $12," Ixch'umil said. "The table runner is $7."

Over the next few days Anne and Ixch'umil talked. Anne watched her weave, she visited her home, and she tried weaving herself. She also visited textile markets around the city. On her last day of class, she purchased a table runner and a basic *tzätz* from Ixch'umil. She later told me that she decided the *maracador* style was too expensive. I took the opportunity to ask her about her choices. "Are these authentic, authentically Mayan?" "Of course," she answered. "Why?" I asked. "Well, these weavings are made by Ixch'umil and they are woven on a backstrap loom. That makes it authentic." Her explanation illustrates Bruner's point: "No longer is authenticity a property inherent in an object, forever fixed in time; it is . . . a struggle, a social process, in which competing interests argue for their own interpretations" (1994, 408).

In the negotiations between Ixch'umil and Anne, it is important to recognize that they do not share a common vocabulary. To clarify, I am not talking about technical, analytical definitions of authenticity, but rather the vernacular uses and the miscommunications that arise in cross-cultural exchanges between Maya vendors and foreign tourists. Although Anne used phrases like, *típicos auténticos* and *tejidos tradicionales* in her conversations with Ixch'umil, Ixch'umil only used them when Anne questioned the textiles' veracity or authenticity. Even then, she responded with phrases that avoided *traditional* and *authentic*: *tejidos verdaderos*, in Spanish, or *ri kan qitzij kem*, in Kaqchikel Maya, both meaning "'true' or 'real' weavings." For Anne, part of what makes a Maya textile authentic is that it is traditional, that it embodies a cultural essence based on histories, practices, places, and people deemed Mayas, yet defined exteriorly by non-Maya experts or culture brokers, be they scholars or tourism industry specialists. When Ixch'umil and I conversed in Kaqchikel about textiles and weaving, she never used the phrase "*ri kan qitzij kem* (true/real weavings)," but instead said *qakem* (our weavings) or just *kem* (weaving, textile). She also tended to be more specific, naming the items *po't* (blouse), *su't* (shawl), *uq* (skirt), *pa's* (sash), but not qualifying them as real, true, traditional, or authentic.

The struggle over authenticity to which Bruner refers is complicated for two culturally different people negotiating across distinctive worldviews. Ixch'umil, like many of the vendors and weavers (see Little 2004), tries to educate tourists about textiles and weaving, emphasizing that the *process* is a key factor in it being an indigenous practice. Tourists are left to decide for themselves whether a textile or weaving itself is authentic or traditional.

INTERPRETING / PERFORMING / THINKING AUTHENTICITY

The exchanges between weavers and tourists are performances that are repeated throughout the Guatemalan highlands. These are certainly staged performances, but I am hesitant to categorize them as staged authenticity. MacCannell (1973) argues that staged authenticity is that place between Goffman's (1959) front-and-back dichotomy. In tourism, MacCannell explains, the place of staged authenticity is where tourists are led to believe they are getting glimpses of and participating in the Goffmanian back stage, the private place. Of course, this is a constructed space for tourists. According to MacCannell, "It is a staged back region, a kind of living museum for which we have no analytical terms" (1973, 596).

Bruner uses Maasai tourism performances to critically explore MacCannell's concept, explaining, "MacCannell sees modern tourists as on a quest for authenticity, which is frequently presented to them as 'staged authenticity,' a false front that masks the real back stage to which they do not have access" (2001, 898). In Bruner's final example, official performers leave the stage and mingle with tourists. I understand his argument to be that authenticity is not salient and is not really the tourists' concern (2005b). They are having fun and appreciate a good performance. I think this is what tourists are doing with Maya weavers and why they do not question the authenticity of weaving as a process.

It is important to emphasize that Ixch'umil's performance is not as Bruner describes for Maasai dancers: "The entire performance was produced to achieve tourist realism, an ambience of authenticity, and the appearance of the real" (2001, 885). Weavers would be insulted if scholars such as myself and tourists considered their public weaving to create "an ambience of authenticity" and to have "the appearance of the real." Ixch'umil and other Mayas are really weaving, and tourists are interested in that. Delineating these spaces in such ways and locating them as staged authenticity would be to construct an artificial space that is out of sync with Maya conceptualizations of what is "true" or "real" (*ri kan qitzij*) and what are everyday practices of weaving and vending. By contrast, although tourists may call these authentic practices, they are more content to participate, or play, by trying to weave and by negotiating with vendors. Bruner notes this playfulness in his research on tours in Kenya, where some tourists "are willing to dance with the Maasai and joke with them, and they are not that fastidious about authenticity" (2001, 899). Tourists did not question the realness of weavers like Ixkem, or Ixch'umil, as Bruner witnessed tourists in Kenya asking, "Are these Maasai for real?" (2001, 901).

It could be that tourists, like Anne, the Spanish student, think that weaving and market negotiations are too obviously authentic to question them. But authenticity can matter to them when they examine and purchase textiles. How they understand the textiles is shaped by their awareness that they are in an exchange in which vendors try to make sales of items of which the tourists have limited knowledge. In instances where weavers gave gifts of textiles to tourists, the veracity of the textiles was not questioned, and tourists generally considered the items "authentic," if authenticity came up at all. Hence, the performance—weaving or market negotiations—could be considered unquestionably authentic, while the textile as a commoditized tourism item is questionably authentic.

MacCannell writes, "once tourists have entered touristic space, there is no way out for them so long as they press their search for authenticity" (1973, 601). This applies to the interactions between Ixch'umil and Anne and between Ixkem and the American client. Anne abandoned some of her notions about what is Maya, especially those tied to aesthetics and materiality, and expanded them to include practice. By contrast, the American client did not accept what can be considered Ixkem's rationale for claiming authenticity, but he did acquiesce to her economic demands. Anne stopped worrying about authenticity, had fun, and purchased something from a weaver whom she got to know personally.

Irrespective of Bruner's and MacCannell's debate, I suspect that there are some tourists who will always be looking for the back stage, even in the case of the Ixch'umil's performance, where there is no back stage—as an actual physically manifested place or as a conceptual space. Yes, there is a performative space, but for her it is without a front stage or back stage. She even weaves while students are in class and not able to engage her, telling me that she has to be productive, (*k'o chi ninb'än nusamaj*). Ixch'umil's slow multiday performance changed Anne's sense of the authentic. In the end, she moved from having a quest to find an authentic Maya textile, to having a social relationship that, in turn, reformed her concept of authenticity for Maya textiles so that it conformed with Ixch'umil's position that weaving processes constitute a textile's authentic and traditional value, not some exterior aesthetic or moral value.

CONCLUSION

The work of Bruner and many others (Gable and Handler 1996; 2005; Cohen and Cohen 2012; Timothy and Boyd 2006; Olsen 2002; Wang 1999) illustrates that debates about authenticity are not without political tensions, can be prone to confusion, and are embedded in power relations. In Guatemala such power relations are embedded in the interactions between weaver-vendors and tourists. The examples in this chapter are drawn from my twenty-five-plus years of ethnographic research with Guatemalan Mayas. I have used these ethnographic situations to explore authenticity as socially constructed coproduction. Rather than focus on authenticity via the lens of Western tourists, I have argued this point from the perspective of those hosting and performing for tourists. In so doing, I have endeavored to explore the Maya side of the social construction of tourism authenticities. Bruner notes this perspective, but it has been lacking or underdeveloped in his discussion of Maasai performances to tourists at Mayers Ranch, as well as in his writings on Bali.

Most tourism research on authenticity has not concentrated on the attitudes of those performing or making items, especially in cross-cultural contexts, though Yujie Zhu (2012) has claimed to focus on performers' perspectives. I suggest that Zhu's analysis is problematic because he never gets to an ethnographic level where performers' attitudes are delineated. Similarly, as Dimitrios Theodossopoulos (2013, 341) notes about Christopher Steiner's (1994, 72) well-quoted exchange between a Western tourist and an African art trader, understanding an object's authenticity across different cultural and other conceptual frameworks is no easy task. African traders use European consumers' notions of authenticity to fabricate their wood carvings, but it is not clear what the African trader thinks *authentic* means. I would argue that this fabrication is the result of a coproduction of authenticity, though the buyer may ultimately be disappointed in learning of this process. Ethnographic inquiry makes it possible to learn what performers, hosts, and artisans think about North American and European tourists who pursue material culture and touristic experiences in Mesoamerica (Brulotte 2012; Castañeda 1996; Feinberg 2003; Little 2004; Schlossberg 2016; W. Wood 2008). My contribution in this chapter illustrates the influence of Bruner's conceptualization of authenticity and, more specifically, the intricate ways in which the meanings of things (here, textiles), cultural performances, and cross-cultural experiences are coproduced. Authenticity may emerge as salient because of these exchanges, but it may not.

Chapter Five

{The Borderzone}

Living in and Reaching beyond the Touristic Borderzone

A View from Cuba

Valerio Simoni

This chapter discusses and expands Bruner's notion of the "touristic borderzone," the "performative space within which tourists and locals meet" (E. Bruner 1999, 461).[1] For Bruner, the touristic borderzone is "a creative space," "a site for the invention of culture on a massive scale," and a location "that anthropology should investigate, not denigrate" (1996b, 159). His conceptualization draws on performance theory and leads him to see tourism "as improvisational theater with the stage located in the borderzone" (2005b, 18–19). While touristic borderzones may be found in any "actual place in the world," Bruner tells us that "what is created there is a cultural imaginary, a fantasy, in itself not a real-life culture but a constructed theatrical one" (2005b, 18). In his reflections, he acknowledges earlier uses of the term *borderzone* by scholars and artists who have theorized it in relation to migration and the US-Mexican border (Anzaldúa 1987; Gómez-Peña 1993; Hicks 1991; Fusco 1989), but departs from them by focusing on what is specific to the touristic realm. Bruner's notion of the borderzone shares aspects of Renato Rosaldo's "borderland" (1989), Mary Louise Pratt's "contact zone" (1992), and Homi Bhabha's "third space" (1994), in that all these terms draw attention to spaces of cross-cultural encounter and their generative potential, without neglecting their power dimension. Bruner's concept is also related to Dean MacCannell's (1992) notion of "empty meeting ground," as "a site of new subjectivities and consciousness where new culture and relationships emerge" (E. Bruner 2005b, 18).

As briefly sketched above, the notion of the touristic borderzone closely fits the research on tourism that has occupied me in the past fourteen years, which is focused on the encounters and relationships that develop between foreign tourists and members of the visited population in Cuba (Simoni

2016b). While I have not dealt explicitly with Bruner's notion of the borderzone in my previous publications, this chapter gives me the opportunity to clarify its pertinence and scope in the light of a very different ethnographic example, one that departs significantly from those that Bruner addressed in his work. While I find the touristic borderzone a very productive concept for analysis of the encounters that happen in tourism, the empirical research I carried out in Cuba complements and expands on what Bruner uncovered and theorized on the basis of his own fieldwork material. Skeptical of grand theory and overgeneralization (E. Bruner 2001, 881), and a staunch advocate of ethnography-based analysis, Bruner is the first to acknowledge the limits of the material on which he draws his theoretical insights. Spanning a wide variety of research contexts—from the United States to Kenya, Sumatra, and Ghana—his empirical work is still mainly focused on "American upscale packaged tours" and what he calls "ethnographic tourism" (2005b, 258n8). Bruner recognizes that "this is just one form of tourism" and that objectives and outcomes of tourism may differ from what he found (2005b, 258n8). By looking closely at some of these different objectives and outcomes, I develop Bruner's insights, feeding back on and extending the reach and implications of his original theorizations.

As Bruner so eloquently advocates, tourism needs to be taken seriously by anthropologists. In this chapter, I take seriously those instances in which the protagonists of tourism strive to move beyond the touristic borderzone. To illuminate the touristic borderzone—what it brings into play and the effects it has—I argue that one also has to examine people's efforts to reach beyond it, what these efforts mean to them, and the far-reaching implications these can have. I do not equate reaching beyond the touristic borderzone with what MacCannell (1976) theorizes as the quest for authenticity; I am not "looking elsewhere, over the shoulder or around the bend" (E. Bruner 1996b, 176) for a supposed original, and I do not imply that what happens in the touristic borderzone is a fake performance that does not merit attention. Rather, by "reaching beyond" I evoke, in the title of this chapter and elsewhere, my research participants' efforts to propel relationships that originated in the borderzone toward new horizons of possibility that exceed tourism's temporal and spatial frame and that deeply transform their lives beyond the tourism realm. This new shared reality, when realized, is to replace that of tourists going back to the same everyday life in their countries and of Cubans reintegrating into their home community after the touristic encounter.

For Bruner, "the tourists come from distant places and the natives come out of their communities to coexist in a specially constructed locality, a performance space that I call a touristic borderzone" (1999, 475). The situation is not so different in the Cuban context in which I carried out research, where the initial divide and distance between the tourist and the native tends to be sharp and clear-cut, also due to the specific features of tourism development in Cuba, whose institutional design pushes foreign tourists and Cubans apart (Simoni 2016b). The consequence of such an initial divide on the nature of tourist-Cuban encounters and where these may lead, however, can be markedly different from the cases analyzed by Bruner.[2] The analysis I develop here problematizes Bruner's vision of the touristic borderzone as a situated event whose resolution sees "tourists return to their hotels and the locals to their homes and families, each to their own spaces" (1999, 475). By unsettling this scenario, the ethnographic material I present also calls into question tourists' and natives' disjunctive perceptions of the borderzone as a space "of leisure and exoticization" for the former, and a "site of work and cash income" for the latter (E. Bruner 1996b, 158).

Most ethnographic examples on which I draw in this chapter come from my monograph on tourism and informal encounters in Cuba (Simoni 2016b). Here I use this material to establish a productive dialogue with Bruner's work, and to discuss and further advance some of its theorization. The findings are based on eighteen months of ethnographic research undertaken in

Figure 5.1. Tourists and Cubans socializing on the Malecón, Havana.
Photo credit: Valerio Simoni.

Cuba between 2005 and 2019, mainly in the city of Havana, the beach resort of Playas del Este (half an hour drive east of the capital), and the rural town of Viñales (approximately two hundred kilometers west of the capital). In these three tourism locations, I carried out interviews and participant observation with tourists and Cuban men and women, focusing on their encounters and on the various types of relationships they established. The research focused on interactions taking place at the margins of the formal tourism sector, which I conceptualize as "informal encounters" (Simoni 2016b). These encounters, characterized by high levels of ambiguity and reciprocal manipulation, led to the renegotiation of a range of relational idioms, among which are those of commerce, hospitality, friendship, love, and sexuality. I first focus on the borderzone of informal touristic encounters in Cuba, then move to a consideration of its limits, and finally to Cubans' efforts to reach beyond it, to move from "performance time" (Bruner and Kirshenblatt-Gimblett 1994, 463) to what my research participants could consider "real life," a life to be achieved through and with tourists.

LIVING IN THE BORDERZONE

The touristic borderzone is a creative site in which culture is performed, but one should not forget that the prevailing scripts, images, and metanarratives that come into play in it are usually part of much larger "tourist tales" (Bruner and Kirshenblatt-Gimblett 1994, 467) with a history of their own, a history in which colonialism and domination often also feature. In the case of the Maasai performances Bruner studied in Kenya, for instance, it was clear that colonialism had played a key role in shaping their global image as "male warriors." Bruner raises important questions about the extent of control the Maasai had over the images by which they were represented and has suggested that they were by and large "willing to play into the stereotypical colonial image of themselves to please their clients, the foreign tourists" (2001, 896). There is a tendency in tourism promotion, which thrives on creating value out of cultural difference, to reify the specificity of the people living in a given destination—their "typical" traits. Bruner reviews some examples of this, including representations of "the Pueblo as female potters, . . . the Balinese as people who live in a magical world of dance and drama, . . . and the Tahitians as representatives of South Seas sensuality" (2001, 881). Cuba is no exception to this global trend.

In terms of Cubans' "defining characteristic[s]" (E. Bruner 2001, 881) as they appear in tourism promotion material, we can identify hospitality, kindness, and joyfulness, with the island's inhabitants described as "amiable and warm," "living out of smiles, dances, and music," and displaying "a lascivious and languorous joy" and a "passion for life" (Michel 1998, 276–277).[3] Prolonging these semantic registers is the emphasis on the exuberant, passionate, sensual character of Cuban men and women, with Cuba made out to be, as one guidebook puts it, "probably the most sensual country on earth, not to say sexual."[4] Echoed in several mainstream tourism publications on Cuba, these images interlock the exuberance, inclination for partying, and diffuse sexual permissiveness that supposedly characterizes Cubans. A number of scholars writing about recent developments in tourism in Cuba have underlined continuities between contemporary sexualized images of "hot" Cuban people and the slave and colonial past of the island (Fusco 1997; Kummels 2005; Sánchez Taylor 2000). Prolonging the (hyper)sexualization of Afro-Cubans (Allen 2007; N. Fernandez 1999; Roland 2011), the stereotype of the "hot" Cuban is nowadays reactualized in a culturalist and nationalist vein in interactions with tourists, and potentially applied to all Cubans (Simoni 2013).

Not unlike Bruner's Maasai, who were willing to play into stereotypical images of themselves if it could help them benefit from interactions with tourists, Cuban men and women's reputation as uniquely "hot" was frequently actualized in the borderzone. Cuban people I interacted with banked on the image the country had acquired on the global tourism market and did not hesitate to praise their amazing sexual skills to foreign visitors. "What? You've never tried [sex with] a Cuban? You don't know what you're missing, my friend"[5] could be the provocative remark, often voiced with a touch of humor and accompanied by a smile or an innuendo-charged wink. "Once you make love with a Cuban, you don't backtrack!" was another classic. These playful remarks could be addressed lightheartedly and could break the ice and bring about an informal atmosphere of complicity and intimacy with tourists. Far from always being innocent, such comments could also test the tourists' interest in sexual relationships with Cuban people, foreshadowing intimate relationships between the protagonists involved. Tourists could subsequently be encouraged to "let go" and indulge in the exceptional sensuality and sexuality of the Cuban Other. If they had really come to Cuba to discover another country and another culture—as normative tourism discourses prescribe—then they had to experience first-hand one of its quintessential typicalities, the self-professed "hot" Cuban would argue (Simoni 2013).

As I came to understand from Cubans and tourists, men and women alike, the way a first connection between them was established played a fundamental role in shaping prospective encounters and relationships. My Cuban friend Fernando,[6] who had been in the informal tourism business for a decade when we first met in 2006, told me that *entradas* (openings) were the most difficult part of relating with a tourist, and his remarks were echoed by many other

Cuban research participants. Once a Cuban had managed to get the tourist's attention and start an interaction on a positive note, interesting and profitable possibilities were likely to open up. As a frequent object of such *entradas* myself, I became more and more familiar with people's techniques, and I noticed that the more original and unusual the initial approach strategy, the likelier the possibility of capturing the visitors' attention. People's ability to decode tourists' behavior and detect their potential interests helped tailor approaches, and devising openings that fit creatively into such interests was a smart way of establishing a connection. What seemed very important was to avoid any hint of premeditation or suggestion of being a full-time tourist hustler. *Entradas* could be designed precisely not to give the impression of a well-scripted, predictable, hustler-type approach, and to be seen as genuine attempts to treat tourists "as persons" (versus gullible "marks")—as individuals with specific personalities, interests, and agendas (Simoni 2014).

The wider the range of assets and competencies Cubans could mobilize in dealing with tourists, the better their chances of establishing and lending continuity to relationships with them. Adaptability became a considerable asset, and it was precisely the importance of such "chameleon-like qualities" (Bowman 1996, 90)[7] that Jorge, a Cuban man in his forties whose turf was in Old Havana, made clear to me in the course of a long conversation we had one afternoon. Describing his informal tourism-related activities, Jorge told me that he could quickly assess what kind of tourist he was dealing with, and this informed the way he would present himself: "If he is cultured, I am cultured. If he is delinquent, I am delinquent." As Jorge indicated, one could easily reinvent one's persona to pique visitors' interest and hold it. Accordingly, the same person could perform a university student for certain tourists (or in his case, given his age, a university professor), a salsa teacher for others, and a cigar manufacturer for others still.

Visitors too could playfully enact different personae to give a twist to their encounters. Tourists could pretend to be experts on Cuba who had spent a lot of time on the island to show that they were not novices and could not easily be deceived. Some went as far as saying they were Cubans themselves, though such claims were generally playful and ironic in tone and quickly unmasked. They might also pass themselves off as rich businesspeople or as poor students with no extra money to spend, enacting "ritual inversions" (Graburn 1983a). They could say they were married when that was not the case or claim a different sexual orientation to avoid Cubans' sexual advances, or become bachelors who, without families back home to account for, were up for sexual experimentation.

Other anthropological work on tourism has also uncovered these performative, theatrical, and utopian dimensions of touristic encounters: from the early research of Nelson Graburn on tourism's potential for self-creation and re-creation (1983a); to Andrew Causey's remarks on "utopic space"—space that tourism offers people to "explore possible ways of being . . . between reality and unrealizable desires" (2003, 167); to David Picard's (2011, 41) reflection on the "kaleidoscope of possible roles and existences, of selves" potentiated by encounters happening through tourism; and to Naomi Leite's (2017, 5) analysis of "how fleeting interactions [with tourists] allow different expressions of self, and hence different experiences of belonging or exclusion, to come to the fore." Michel Foucault's notion of "heterotopia" (1986) has similarly been used to theorize the qualities of "heterogeneous tourist spaces" (as opposed to "enclavic" ones) and their potential to act as "places for exploration and self-expression" (Edensor 1998, 43).

Bruner's notion of the touristic borderzone is a powerful analytical tool to recognize and highlight these creative and generative qualities of the touristic encounter. Foregrounding the theatrical and performative dimension of these encounters, the touristic borderzone draws our attention to the continuous invention and reinvention of selves and cultures that tourism potentiates, and, I would add, of human relationships and how we imagine them. In connection to this last point, my work shows how touristic encounters are a site in which the relational idioms of hospitality, commerce, friendship, seduction, sexuality, and love are brought into play and worked over in creative ways, leading to the transformation of their meanings, expressions, and implications (Simoni 2016a; 2016b).

THE LIMITS OF THE BORDERZONE

"I want to know if what they [tourists] are telling me is true or not," Dennis told me one day. Determined not to come off as a fool in conversations with foreigners, this Cuban man in his early thirties asked me specific questions about life abroad, things he wanted to know so he could engage with tourists from a similar standpoint and level of knowledge. His aspiration was to base his exchange on shared cultural competencies and wisdom. Grasping multiple points of view, understanding different approaches to tourism and tourists, attending to the subtleties and sensibilities of relating to people who differ in terms of their origins, age, gender, socioeconomic status, interests and more: the practice of informally engaging with tourists is seen by some as *un arte*—an art of communicating and dealing with people, of sparking interest and developing relationships, a positive body of knowledge about humans and human relationships.

Here we rejoin the considerations of Jill Forshee (1999), Susan Frohlick (2007), Leite (2017), and Noel Salazar (2010), who in different touristic contexts—studying pedicab drivers in Yogyakarta, Indonesia (Forshee), sexual and intimate relationships between female tourists and local men in Caribbean Costa Rica (Frohlick), interactions between self-identified descendants of Inquisition-era Jews and international Jewish heritage tourists in Portugal (Leite), and tourist guides in Yogyakarta and Arusha, Tanzania (Salazar)—have likewise highlighted the "cumulative knowledge" (Frohlick 2007, 149) of tourists' preferences, imaginaries and desires, and the peculiar cosmopolitanisms that members of a visited population can develop through their engagements in the tourism realm. In all these cases, what we have are clear illustrations of the touristic borderzone's potential to act as a key site for the invention and transformation of culture (E. Bruner 2005a).

Most of the Cuban men and women I interacted with appreciated being able to learn new things from visitors and explain to them in turn about "their" Cuba, a Cuba they saw as more real and authentic than the one depicted in tourism brochures. My interlocutors valued the possibility of making a living out of their interactions with foreign visitors, thanks to what they managed to sell them informally, the commissions they could gather in the tourist bars and restaurants, or the presents they got in return for their help and guidance. This, however, was not all that mattered for many of them. The touristic borderzone, with all its potential for creativity, performance, and value creation, could be a lively and even fun place to be for some, and a place to eke out an existence, but its frame could also represent a limit for those who aspired to generate something more substantial and truly transformative from their relations with tourists.

Bruner's view of tourism as "improvisational theatre," with the borderzone as its stage (2005b, 18–19), is analytically useful in that it helps us depart from the more static views of "hosts and guests" that prevailed in earlier theorizations of tourism (V. Smith 1977). The notion of the borderzone opens our eyes to tourism's potential for generating new intersubjective meanings between tourists and members of the visited population, enabling us to recognize that the stage these protagonists share is a key site in which culture is produced and invented, and one that anthropologists should take very seriously in their analyses (E. Bruner 1999, 461–62). The improvisational theater taking place in Bruner's borderzone is an event that is clearly demarcated in space and time. Before, after, and, to a large extent, also during this event, there remains a divide between tourists and locals. There may be moments, Bruner tells us, in which "boundaries" become elusive and less rigid—in which "tourists and natives . . . move into each other's space" (2001, 896). Bruner gives some examples, such as the case in which a Maasai guide announced to a tour member he wanted to marry her daughter. Alongside these exceptional moments, the borderzone remains "an empty stage waiting for performance time" (1996b, 158), lending credit to the idea that it is better not confuse "performance time with real life"—or "tourist time and life time" (Bruner and Kirshenblatt-Gimblett 1994, 463).

Had he been confronted with the situations I describe in this chapter, events that appeared exceptional in Bruner's research sites in terms, for instance, of tourists and natives moving into each other's space—a "transgression" of a prevailing "spatial pattern" (E. Bruner 2001, 895–96, referring to the work of Michel de Certeau 1984)—might have seemed more common, and at times even the key stake of some touristic encounters. This, in turn, might have led him to pay more attention to the potential meanings and implications of such moves and "transgressions"—to the importance that reaching together beyond the limits of the borderzone, and moving relations from "performance time" to "real life," could acquire for the people involved.

Embracing the theater-like mode of the touristic borderzone, tourists I met in Cuba would get along happily with friendly Cubans who asked them to join and spend some time with them, knowing all the while that this was just part of "tourist time," a performance moment after which participants would reintegrate into the communities to which they belonged in "real life." In the case of intimate sexual encounters, several tourist men were keen to engage "hot" (*caliente*) Cubans, to explore their reputed sexual ardency and passion, while clearly framing such experiences in the exceptional time of the holiday, and careful not to make them overflow into their "real lives" back home. "*Moglie e buoi dei paesi tuoi*" (wife and ox from your own land) was the Italian proverb offered by two Italian men in their sixties who had been regular visitors to Cuba for many years and were self-proclaimed experts in Cuban women. Their reasoning implied that while it was ok to enjoy sex with "hot" Cuban women in the touristic borderzone, for a serious decision like marriage it was wiser to choose someone with your own background, from your own country—someone you could trust. There was no point in taking seriously Cuban women's declarations of love, the men argued, as these were just a performance and likely an expression of typical Cuban trickery, a way of seducing the tourist into a "fake" relationship for purely material interests, chief among them the possibility of migrating via marriage to the tourist's country of residence.

Love in these relations was seen at best as a tool to play with in a theatrical mode, in a game that could be detrimental to the Cuban women who were unwillingly caught on a stage that was not of their own making. With no explicit agreement or common understanding of the frame in which relationships were taking place, what was one party's "performance time" could be the other's "real life," creating a situation rife with possibilities for playing with power, ambiguity, and misunderstandings (Simoni 2016b). In the most cynical tourist discourses I heard on the matter, the place for romance in relations with Cubans was as a tactic to spark false hopes of "love and migration," and thus capitalize on the ambiguities of the situation to lure Cuban women into having sex for no payment at all.

In the case of Cubans' declarations of friendship, tourists could deem them as well to be mere make-believe, and thus not to be taken too seriously. Friendship claims could even become counterproductive when seen as a way of concealing instrumental agendas. Hazel Tucker (2001, 880) describes a scenario in the Turkish village of Göreme, in which tourists often became "suspicious of the perceived over-friendliness of salesmen and waiters." Particularly in the touristic areas of Havana, visitors easily tired of over-friendly approaches, marking the Cubans that addressed them this way as the typical hustlers tourists should avoid. Tourists could also decide to go along with a friendship claim while playing "as if," maintaining a playful distance in a stance akin to that of the "post-tourist" (Feifer 1985, 267; Urry 1990, 11), which sees acutely self-reflexive tourists play with normative scripts and roles and become complicit in tourism-related *mises-en-scène*.

Bruner often described this type of interaction insightfully. His analysis of tourists' appreciation of the Maasai performance at the Kichwa Tembo Tented Camp in Kenya identifies this as a theatrical scene that works in the mode of "make believe" (2001, 890–891), and that echoes the postmodern ethnographic canon of "juxtapositions, pastiche, and functional inconsistency" (2001, 899). In this "surreal theatre" (2005a, 96), "post-tourists are willing to dance with the Maasai and joke with them, and they are not that fastidious about authenticity" (2001, 899). Although "the distance is narrowed between us and them, subject and object, tourist and native" (2001, 893), the interaction remains firmly anchored in a well-delineated, momentary touristic borderzone, and before long the tourists and the Maasai reintegrate into the separate spaces and the communities to which they belong. Once again, the limits of the touristic borderzone and of what one could ask and expect of it are clear. We may argue that it is precisely because they know that this is "tourist time" that visitors get involved and relate with locals in the first place. For several of my Cuban interlocutors, however, the limits that the frame of the touristic borderzone as a space of performance posed to a more serious, full-fledged involvement in the relational idioms they tried to bring about with tourists became a major obstacle they wished to surpass. To surpass that obstacle, both tourists and Cubans had to reach together beyond the borderzone.

REACHING BEYOND THE BORDERZONE

In her work on tourism in Cuba, Amalia Cabezas shows how resort workers employed in hospitality organizations that encourage "friendliness, subservience, and flirting" (2006, 515) with tourist clients blur the line between the behavior suggested by hotel management and the pursuit of their own agendas. Being friendly to tourists and flirting with them as a service provider fits the performative dimension of Bruner's touristic borderzone, made of complicit interactions with tourists under the sign of make-believe and playing "as if." In my own work, I have similarly described how Cuban women could playfully seduce tourist men on the dance floor, arousing them with sensual dance moves presented as "typically Cuban" (Simoni 2016b). Playful dancing performance, however, could take a more serious twist and flow into seduction away from the dance floor, sexual relations, and eventually a more sustained love affair—the latter being what some Cuban women could precisely be hoping to achieve. In Cabezas's case, we also see resort workers seeking opportunities to cultivate intimacy with hotel guests and to establish "relationships that create long term obligations and commitment" (2006, 516). The potential for romance and marriage with tourists, loaded with opportunities to leave the country, could thus become the most attractive prospect of employment in all-inclusive resorts (Cabezas 2006). As strong personal bonds are forged between Cuban employees and foreign tourists, the initial staged performance of intimacy that hotel managers call for, aligned to the broader image of Cubans as flirtatious and *caliente*, shifts into another realm that breaks down the client/worker divide, opening other relational possibilities for the protagonists.

Not unlike some of the resort workers interviewed by Cabezas, several of my Cuban interlocutors aimed, in their engagements with tourists, to tear down the divides separating them, not just for a short-lived interaction in a clearly delimited touristic borderzone, but for an indefinite term, prompting a switch from "performance time" to "life time" (Bruner and Kirshenblatt-Gimblett 1994, 465). The drive for engaging in the borderzone, for some, went beyond making a living and establishing beneficial, possibly gratifying connections with tourists. The more ambitious vision was to radically alter their existence and realize deeper aspirations, among them finding "true" love and a "normal" place to live, and infusing new hope, meaning, and prospects into their lives—things that people felt they could not realize in contemporary Cuba. For Cubans who projected these expectations onto their engagements with tourists and believed in their far-reaching potential, the visitors' skepticism, distance, and playfulness toward letting go, opening up, and committing to relationships could prove very disheartening and disappointing, hampering the possibility of developing more long-term, open-ended bonds.

After more than a decade relating with tourist women, Umberto, a Cuban man in his thirties with whom I spent several evenings discussing relationships with them, was now in search of a "serious person" (*una persona seria*), a foreigner who would take seriously the possibility of having a long-term love relationship with him. To the *persona seria* he opposed the *bandolera* ("bandit"): someone who was just looking for sex, fun, and short-term gratification in the exceptional time of the holiday without making any longer-term commitment—as many tourists seemed more inclined to do nowadays, or so his criticism went. In their claims of friendship and love, many Cuban men and women I met wanted to be taken seriously. They were ready to put their entire lives into the equation, in the hope that a long-lasting relationship, and perhaps even a life together, could come out of it (Simoni 2016a). To render this scenario possible, however, it was necessary for them and their foreign friends and partners to burst the bubble of a temporary touristic borderzone, to move from the realm of "performance" to that of "real life."[8]

Many of my Cuban interlocutors spent most of their days trying to interact with tourists: the touristic borderzone was a key site of their everyday life. But to say this does not tell us much about what this site and this life looked like. As Bruner's conceptualization indicates, the touristic borderzone is indeed a site of work, or of "play-labor" as Allen puts it (2007, 189); of theatrical relations with tourists; of improvisation and self-creation—but it is also the site that many wish to overcome through relations with tourists and their generative potential. Cuban men and women thus devoted much effort to teasing out relationships from the overarching realm of the touristic, and to extricating the protagonists involved from their overdetermined identifications as exemplars of general tourist and Cuban populations (Simoni 2014).

"For me you are not a tourist. You are a person, a human being!" In the course of my fieldwork in Cuba, I repeatedly heard this statement from Cubans engaging with foreign tourists. The first temptation may be to treat this as a typical flattering remark designed to reassure skeptical tourists, to try to convince them that they have nothing to worry about because they will not be (mis)treated as "mere tourists," but I came to understand that there was more one could read into this claim. The remark was an invitation to break preestablished boundaries and typifications (Cuban/tourist, host/guest, us/them), to recognize the partners involved for who they were as persons, with all their peculiarities, idiosyncrasies, and unique circumstances. Of course, it was also a call for tourists to reciprocate and do the same with their Cuban partners, and to commit to full-fledged relationships with them—relationships that would respect the integrity and complexity of the other, without reducing the other to a type, a summative account, a partial representation and example of a more general category (see Simoni 2014; Throop 2014; Leite 2017). We can thus read this remark as a demand for openness, for partaking in a world as yet unwritten and unscripted, full of generative potential in terms of ways of being and doing things together.

"Occasionally tourists and natives do move into each other's space," Bruner tells us, "as boundaries are not absolutely rigid and there are transgressions and resistance" (2005b, 19). The demand that I and many other tourists heard from our Cuban interlocutors, however, was not so much that we step into their world. This could arguably be a step in the relational process Cubans strived to conjure, a way for us to realize their "real life" in Cuba, beyond what they saw as the superficial image that was presented to tourists in more official circuits. Nor was it a demand that Cubans simply jump into the tourists' reality without changing it. Rather, it was a call for tourists and Cubans to step out of their typification as tourist and Cuban and to be ready to trace new shared paths into the future—a relationship and a future placed under the sign of friendship, or of love, that could ultimately result in a life together.

The space people had in mind for such a life together was generally outside of Cuba, in the foreigners' countries of residence, where life was deemed more "normal" and they could escape the harsh living conditions that still prevailed on the island. Several of my Cuban interlocutors saw Cuba as a country of exception, as a place isolated from the rest of world, where material needs were hard to satisfy and instrumental forms of sociality were on the rise (see Fosado 2005; Lundgren 2011). Their aspiration was to leave such a place behind and to be able to live a "normal" life somewhere else.[9] As I have elaborated elsewhere (Simoni 2015; 2016a; 2016b), such longing for "normalcy" (Patico 2009)[10] was mediated by idealized idioms of love and friendship, "normative intimacies" (Berlant 2011) embodying the ideal of a "normal life." This was the kind of life that many Cubans I met imagined tourists to have back in their countries, and which they aspired to share with the foreign friends and lovers they encountered in the borderzone. Their demand was for these friends and lovers to step out of the performance time of the borderzone and take seriously their invitations to shape together a new "real life."[11] Such real life was not MacCannell's back stage, the "authentic" Cuba beyond the borderzone, but a life yet to come, in a better, more "normal" elsewhere. With the recent changes in economic policies and the opening of some entrepreneurial opportunities in Cuba, which most of my interlocutors viewed positively, the site of a projected future could sometimes be located on the island itself, but an island to be lived differently—beyond the exceptional, enduring states of crisis and scarcity they experienced and criticized, and beyond the touristic borderzone too.

CONCLUSION

The analytical challenge that these examples and reflections pose to us parallels to some extent the one outlined by Charles Piot in his thoughts on the current moment of history in West Africa and the demands it makes on anthropologists "for a different set of critical theories" (Piot 2010, 8). Piot's ethnographic work in Togo leads him to characterize the situation in the country, and in West Africa in general, as exemplary of a new "post-postcolonial" terrain (2010, 16) that "postcolonial theory's focus on hybrid culture," for instance, "stops short in analyzing." As Piot makes clear, this is not to say "that cultural hybridity *per se* has vanished, here or elsewhere," but rather to suggest that the "hybridizing" typical of the colonial and postcolonial periods "is no longer desired and is on its way out" (2010, 172n8). My fieldwork in Cuba confronts me with a similar situation. On the one hand, touristic borderzones that share the features identified by Bruner are everywhere to be seen, as are the cultural inventiveness and resourceful hybridization these borderzones bring about. At the same time, if we follow the aspirations and desires of the Cuban protagonists who live and make a living in such borderzones, we are forced to realize that their desires and aspirations push the imagination somewhere else, beyond appreciation of the hybridity of the touristic borderzone, and more toward what Lauren Berlant (2011, 167) calls "the fantasy life of normativity," a normativity that seeks roots in a "normal" place where a "real" existence can be carved out.

In analyzing the processes, claims, and demands that my ethnography brings to light, the touristic borderzone, as conceptualized by Bruner, retains a key importance. At the risk of overly simplifying, we may argue that what first appears to be my Cuban research participants' chief everyday life among their "Cuban community"—the "life" time Bruner opposes in some of his examples to the "performance" time of the borderzone (for example, the Maasai after "the picture" has closed for the season at Mayers Ranch; see Bruner and Kirshenblatt-Gimblett 1994, 461)—was for many a site they wished to leave behind, a reality that had lost much of its interest and value, and where they didn't believe their aspirations could be realized. For some, the touristic borderzone in which they invested so much time and energy had become a more valued everyday, and made for a life that was heavily marked by performance, self-creation, and the (re)invention of culture in interactions with tourists. But the touristic borderzone was not the end of the journey on which their eyes were set.

"*Esta es una vida loca*"—"this is a crazy life"—a couple of Cuban friends told me in reference to living in the informal tourism realm and their day-today contact and relations with tourists, with its experiences of friendship and romance, and the frequently frustrated hopes for establishing longer-term relationships that could open up new prospects and scenarios. The borderzone had become for them a site of much stress and uncertainty, where a living could still be eked out but at the cost of both economic and existential stability, where deception—about who they were, who the tourist was, and what the encounter was all about—was always around the corner, and where transitory moments of interaction were tainted by an aura of exceptionalism, elusiveness, playfulness (in the postmodern mode), and potential fakeness (MacCannell's trope). For people who shared in the emerging sensibility and moral disposition I describe here, the touristic borderzone didn't match the "real" life they aspired to; it was just a necessary step along a more ambitious trajectory.

In our analyses of the touristic borderzone we must factor in its value in relation to other realms of people's lives, whether their home community or, when home's location itself is destablized, the future realities they aspire to inhabit. Taking into account the analytical richness and versatility of Bruner's touristic borderzone, its open-endedness and ability to encompass a broad range of endeavors and encounters, I wish to draw attention here to the importance of acknowledging its limits, too, especially when these become essential for our research participants, their relation with tourists, and what they hope to make of it. By following our interlocutors as they delineate the scope of the borderzone, we will gain further insights into the specific features of this important realm of human experience and interaction, and we will better understand what distinguishes it from what surrounds it and what may stem from it, pointing in new directions. We must be ready to account for the linkages between touristic borderzones and other areas of sociocultural change, creativity, and mobility—not to view such other realms as more "real," and thus to reintroduce the simplistic front stage–back stage binary that Bruner so convincingly criticizes, but to understand how our research participants themselves articulate different realms of their lives, assign different value and possibilities to them, and try to move from one to the other.

In the Cuban cases I have presented here, a consideration of the aspirations of Cuban men and women who dwell in the borderzone but aim to reach beyond it enriches the analysis, shedding light not only on what is at stake in touristic encounters, but on what these encounters may be geared toward, on what may come next. By including such aspirations and future orientations in our analytical framework, we may better grasp the way touristic borderzones bring about change, articulate with the past, and are lived in the present.

NOTES

1. This article draws and expands on sections of the book *Tourism and Informal Encounters in Cuba* (Simoni 2016b). I would like to thank the Portuguese Foundation for Science and Technology (Post-Doctoral Grant SFRH/BPD/66483/2009) and the Swiss National Science Foundation (*Ambizione* Fellowship, PZ00P1_147946) for supporting my research and writing. I am indebted to Naomi Leite, Quetzil Castañeda, and Kathleen Adams, the editors of this collection, as well as the book's anonymous reviewer, for their useful suggestions and comments on earlier versions of the text. The chapter would not have been possible without the collaboration of the many tourists and Cuban men and women I worked with in Cuba, and my deepest gratitude goes to them

2. See Simoni (2016b) for a more elaborate description and analysis of *tourist* and *Cuban* relational identifications as actualized in contemporary Cuba, and of the structuring role played by official tourism developments, policies, and policing in strengthening and magnifying tourist/Cuban differences and separations. In the touristic encounters on which I focused, we may argue that the tourist/Cuban divide constituted—initially at least—the prevalent "grammar of distinction" (Comaroff and Comaroff 1997, 25), an overarching and pervasive frame that could easily encompass other distinctions and identifications, and that allotted a key discriminating role to the asymmetry of resources between tourists and Cuban people. As I argue in this chapter and in broader reflections elsewhere (Simoni 2016a; 2016b), however, the possibility of overcoming this divide by bringing about long-term, life-changing forms of relationality, intimacy and togetherness, was essential to many of my research participants.

3. All non-English citations have been translated by the author.

4. *Le guide du Routard: Cuba* (Paris: Hachette, 2007).

5. Direct quotes from research participants have been translated by the author and are based on recollections after the events took place.

6. All personal names and some details in the examples presented in this paper have been changed to protect the anonymity of research participants.

7. The parallel here is with Bowman's (1996, 90) research in a Palestinian tourist market, where he remarks that "street merchants prided themselves on chameleon-like qualities, being able to shift language, religion, politics and even their national identity to suit what they perceived to be the tastes of potential customers."

8. Efforts to transform transitory touristic encounters into long-lasting, life-changing relationships are not unique to the Cuban tourism context and the situations described here. Similar relational dynamics have been described in studies of other destinations around the world, from the pioneering work of Erik Cohen on marginalized "Arab boys" in Israel (1971), to that of Vincanne Adams among Nepalese Sherpas (1992), to the growing body of research on intimate liaisons involving sex and romance between tourists and members of the visited population, which covers cases in Jamaica (Pruitt and LaFont 1995), Thailand (Cohen 1996), Costa Rica (Frohlick 2007), Indonesia (Dahles and Bras 1999), the Dominican Republic (Brennan 2004), Senegal (Salomon 2009), Egypt (Jacobs 2010), and Brazil (Piscitelli 2013).

9. Drawing a parallel with Bruner's reflections on locality and the different meanings that locality as "lived experience" can acquire, we may see my Cuban interlocutors as having a view of their living place similar to the one Bruner ascribes to young Batak toward their village in Sumatra, as "a place from which to escape, a site of lifelong drudgery, poverty, and, for some, a symbol of their failure to position themselves so as to experience the excitement and opportunities available in the larger world" (E. Bruner 1999, 475).

10. In her research on Cuban women's discourses of love for foreign tourists, Dina de Sousa e Santos similarly quotes one of her informants arguing that "people here want to have what is normal to have, simply what any person in the world can have [the world here representing Western countries]" (2009, 422; bracketed passage in original).

11. One may object that what I am referring to here are essentially discourses, ideals, and aspirations, and that inviting tourists to reach beyond the touristic borderzone is different from actually being able to do it. My point here, as I have elaborated more in detail elsewhere (Simoni 2016a), is that we should include in our analyses "the enabling and animating aspects of people's hopes and desires" (Moore 2011, 25). According to Henrietta Moore (2011), it is this "ethical imagination"—"the forms and means . . . through which individuals imagine relationships to themselves and to others"—that ultimately "links human agency to the forms of the possible" (2011, 16, 18). Heeding Moore's call to recognize the importance of "the aspirational character of our relations to others" (2011, 10) is a productive way to bring into the focus of our analyses that which inspires people's actions. It is the generative potential of Cubans' "fantasized relations with others" (2011, 28) to which I draw attention here, as well as to the possibilities for self-formation and transformation it could open up. Drawing on the work of Maïté Maskens and Ruy Blanes (2016, 133), we may argue that Cubans' declarations of love and friendship played "with the gap between reality and fiction as a productive process." These declarations acted, in the utopian genre, as a "call for action" (2016, 134), a call to inhabit such a relational space, to work together in order to bring about such an ideal reality. A similarly productive way to rethink the relationship between ideals and actual practices and engagements can be detected in Rane Willerslev's (in Venkatesan et al. 2011) research on love and the significance of ideals of love among Siberian Yukaghir hunters. Drawing on the notions of "actual" and "virtual," Willerslev uses his ethnographic material to show that "the actual does not exist separately from the virtual" and that "the two dimensions are given as facets of one and the same expression or reality—that is, our actual existence duplicates itself all along with a virtual existence" (Venkatesan et al. 2011, 228). Accordingly, the presence of such virtual ideals should be apprehended as an aspect of the real, as having a significant impact on reality. This is what sustains actual engagements and exchanges, which would be impossible to actualize otherwise. This is what sustained the Cubans' claims of love and friendship, and of a possible life together with their tourist partners, as I am evoking them in this section. Some such claims were taken seriously by tourists, and projects and actual trajectories of life together did emerge from the touristic borderzone.

Chapter Six

{Constructivism}
"I Can Feel Them Now, Even as I Write"
Hiking Yosemite Falls with the Emergent Subjects of Tourism

Sally Ann Ness

> Self and society are always in production, in process; and one of our tasks as anthropologists is to specify how, in concrete instances and in different cultural settings, this shaping and reshaping takes place.
>
> —E. Bruner 1984a, 3

One way to continue to develop the study of tourism as a topic of serious anthropological research is to interrogate the philosophical questions it raises. This chapter focuses on one philosophical question that has been a central concern of Edward Bruner's ethnographic work on tourism: the question of how human meaning is found or made—or shaped, as the quote above characterizes it—in touristic environments. The chapter analyzes tourist activity, specifically hiking, in Yosemite National Park, to illuminate how this question arises and to specify how much agency individuals actually have in relation to its answers. While the Yosemite context differs markedly from the touristic sites Bruner has studied, it nonetheless affords a comparable perspective on human experience as it relates to the creation of both individual and social meaning.

In what follows, I identify Bruner's perspective as pragmatic constructionism. It is pragmatic in its alignment with the pragmaticist philosophy and semiotic theory of Charles Sanders Peirce (Peirce Edition Project 1998). It is constructionist—or constructivist, as Bruner (2005b) uses the term—in the sense of social constructionism. I argue that Bruner's pragmatic constructionism derives from a theoretical orientation that prioritizes performance over both textual and material contexts and thus integrates what are often taken to be opposing philosophical views on how individual, material, and social experiences contribute to the symbolic foundations of human understanding.

As I illustrate in the Yosemite case, and as Bruner's work also demonstrates, such oppositions become illusory when subject formation is considered as creative and emergent, as being in performance, rather than as predetermined.

This chapter is framed as an analytical hike, a metalogue on hiking in Yosemite Valley. At the outset—setting up camp, as it were—I first provide a brief history and description of the Yosemite Falls trail. Once the meta-hiking begins in earnest, I narrate an auto-ethnographic experience that actually took place on the trail in May 2011. This account provides a "concrete instance," as Bruner puts it, for engaging phenomenological, constructionist, and pragmatic semiotic perspectives on meaning-making as they seek to explain productions of self and society. At trail's end I return to Bruner's work, identifying the play of influences that have produced its integrative theoretical approach.

SETTING UP CAMP: CONSTRUCTION OF THE YOSEMITE FALLS TRAIL

The hike to the top of Yosemite Falls is one of the most popular single-day hikes in Yosemite National Park. One of the several "Legendary Trails" leaving from the valley floor, the Falls trail leads to the top of North America's highest waterfall, whose three continuous sections measure 2,425 feet (739 meters) in height (see figure 6.1). From the viewpoint at the Yosemite Falls overlook (see figure 6.4), visitors can peer down at the valley floor on a scene that guidebooks describe as "unforgettable," "worth any amount of hard climbing," "heart-stopping," and "like something out of Disneyland" (Swedo 2011, 46; Wells 1998, 67). A visit to the park—and Yosemite National Park receives roughly four million visitors a year[1]—could hardly be considered complete, particularly for overnight visitors, without a hike at least partway up this trail.

Engineers and laborers built the Yosemite Falls trail shortly after the valley was designated a preservation area in 1864. Begun in 1873 by the master road builder John Conway, the trail was finished in 1877 (Russell 1992, 78).[2] It opened for use during the park's stagecoach era, when access to the valley could be obtained only via horse-drawn vehicles traveling over private toll roads whose primitive condition made the journey both arduous and dangerous (Sargent 2000, 1–6; Johnston 2000). Originally developed to operate on a private, toll-collecting basis, the trail was purchased by the state of California soon after its completion and opened to the public for use free of charge in the early 1880s (Russell 1992, 78). It remains to this day largely unchanged in its design and topography, a monumental feature of the valley landscape dating back to its initial development as a nature tourism destination.

Figure 6.1. Yosemite Falls seen from the visitor's viewpoint on the floor of Yosemite Valley.

Photo credit: Sally Ann Ness.

The Falls trail is considered to be one of the valley's relatively "strenuous" hikes (Swedo 2011, 46). It stretches 3.6 miles long (5.8 km) from the valley floor to the rim of the valley's northern wall, gaining a total of 2,700 feet (823 meters) in elevation. It ascends to the west of Yosemite Falls by an elaborate series of stone switchbacks (see figure 6.2). Rarely does the environment appear conducive to the trail's presence. On the contrary, the granite steps and the marvels they enable visitors to behold evidence the herculean efforts of Conway's engineers and laborers, as well as those of the National Park Service (NPS) seasonal trail crews who have maintained it over the decades.

Progress along the trail can be divided into four stages. The first, very steep, stage covers the first mile (1.6 kilometers), which rises up 800 feet (244 meters) through a wooded area at the base of the northern wall, out of sight of the Falls. This section ends at a small, relatively flat viewing area named Columbia Point (see figure 6.3). At this lookout point, visitors may pause to rest or take pictures of the vistas that become available here for the first time, and possibly turn around and descend if the hike now appears too steep or too long.

The second stage is a shorter, relatively easy, flat section. It moves eastward toward Yosemite Falls for about a mile (1.6 kilometers), gaining only a few hundred feet in elevation. During this stage, Upper Yosemite Falls comes into view for the first time and the entire falls sequence can be viewed at close range. The third and fourth stages are two series of steep stone switchbacks, each about a mile (1.6 kilometers) long and each gaining about 1,000 feet (305 meters) in elevation. They traverse the latter half of the northern wall, up to its rim. At the end of the fourth stage, visitors emerge onto a rocky area where a flat trail guides them the final few yards to the top of the falls (see figure 6.4). From here, they may explore the area near Yosemite Creek and take in the views around and beside Yosemite Falls. Visitors descend to the valley floor via the same route. Guidebooks and NPS literature alike estimate that visitors will spend six to eight hours total if they hike all the way to the top of the Falls trail and then return back down to the valley floor to finish (Medley 2002, 33).

The Falls trail is neither the most renowned nor the most challenging day hike in Yosemite National Park. The hike up Half Dome, for example, is twice as long and gains nearly twice the elevation. However, the Yosemite Falls trail is the hardest hike that the average visitor to the park is likely to attempt. With its trailhead standing just five minutes' walk across the road from the centrally located Yosemite Valley Lodge, and only about a half-mile (0.8 kilometer) from the main visitors' viewpoint at the base of the Falls themselves, it is the most easy-to-find, easy-to-reach trailhead of any of Yosemite's Legendary Trails.

Figure 6.2. Switchbacks section near the top of the Yosemite Falls trail.
Photo credit: Sally Ann Ness.

Figure 6.3. View of Yosemite Valley from Columbia Point lookout.
Photo credit: Sally Ann Ness.

Figure 6.4. The drop-off point at the top of Upper Yosemite Falls.
Photo credit: Sally Ann Ness.

APPROACH: THE TWO WORLDS OF SOCIAL CONSTRUCTIONISM

Given the touristic history and design of the Yosemite Falls trail, the practice of hiking it lends itself well to the study of how tourism shapes social and cultural constructions of both self and place, particularly from what Tomas Pernecky (2012) has termed a "social constructionism" point of view. This perspective, closely associated with Edward Bruner's interpretive approach, focuses on socially and culturally determined productions and transmissions of meaning as they may be evidenced in the activities of individual tourists (Pernecky 2012, 1121–23).[3] Pernecky characterizes the constructionist perspective as a rejection of "objective knowledge," wherein human beings are seen as living in two separate worlds: the first consists of physical and material, or "real," phenomena; and the second consists of socioculturally constructed meanings and ideas (2012, 1124).[4] The Yosemite Falls trail—including its human-designed pathways and its extensive official, commercial, and visitor discourses—exemplifies second-world social and cultural constructions. While the trail's other, "first" world of non-human-made material characteristics is undeniably present and can be hazardous, even fatal, if ignored, when understood from a social constructionist perspective, the trail must be considered symbolically overdetermined by its ideological, economic, technological, and Eurocentric discursive sociocultural constructions alone. At every one of its twists and turns, visitors submit their individual first-world bodies to the trail's second-world social and cultural constructions, enacting and contributing actively to ongoing (re-)constructions of ecotouristic selves.

Elsewhere I have adopted both post-phenomenological and Peircean semiotic perspectives to identify the limits of social constructionism's two-worlds approach with regard to how self–world relationships may form and acquire subjective significance in Yosemite (Ness 2011; 2015; 2016). While constructionism may have much to offer in its focus on established conventions and discourses, the brand of humanist symbolic determinism it espouses goes too far in its claims to prefigure all forms of meaning-making and subject formation that occur in such landscapes. Social constructionism forecloses recognition of the manner in which changing self–world relations may become sources of intelligence and symbolic growth for those caught up in the acts of living through them. Ecological forms of meaning-making, to follow Gregory Bateson's cybernetic model (1972) as well as Tim Ingold's phenomenological approach (2000), may also emerge through direct relations to what social constructionism would identify as first-world or objective characters of the landscape.

To provide a concrete instance that will illuminate these limitations and open a broader array of theoretical and interpretive possibilities, I narrate in the following section an incident that occurred while I was hiking the Yosemite Falls trail in 2011. The incident generated a moment of unanticipated self–world shaping or understanding, a moment in which a sign of the valley landscape was unexpectedly born. The emergence of this sign brought the two worlds of the constructionist perspective into creative and meaningful relation to one another *without* sociocultural conventions playing the sole defining role in that process, thus refuting social constructionism's anthropocentric discursive determinism. However, this case does nothing to undermine—indeed, it is in keeping with—Bruner's interpretive approach. In this regard, the incident foregrounds the ways in which Bruner's ethnographic constructivism is distinctly pragmatic in character. Its emphasis on creativity and performance leads toward a reconciliation of two-world constructionism with realist phenomenological, ecological, and semiotic approaches.

TRAILHEAD: THE FALLS TRAIL HIKING INCIDENT

The incident in question occurred as I was making my way up the fourth section of the Falls trail, although when it happened, I wasn't sure how far along the trail I was. I was accompanied on this occasion by a research assistant, Ariel Bohr, who had hiked with me on a previous attempt in 2009, and Derick Fay, a colleague from University of California at Riverside. Our decision to hike the Falls trail on this calm, sunny mid-May morning was a Plan B alternative we chose when another trail turned out to be closed. Consequently, it was already past 11am by the time we got to the trailhead—a late start. Derick, a fit and experienced hiker, quickly disappeared up the switchbacks to be seen again only at the top. Ariel chose to stay behind with me, preferring what was for her a leisurely pace.

The early phases of the hike went by without mishap. When I reached the point where Ariel and I previously had turned back in 2009 due to snow, I was confident I would reach the top in decent, if not good condition. As best I could gauge, we seemed to be around three-quarters of the way up the trail. In reality, we were not much beyond the halfway point. This miscalculation was a key factor in the incident that was to occur. It served to make the distance seem far longer than it might otherwise have seemed and initiated a growing sense of being unable to read the signs that were intended to keep visitors oriented, motivated, and safe. After this error, I was still on the trail physically—still in its first-world empirical location—but I found myself increasingly off the marks of its second-world sociocultural script.

The incident occurred at around two o'clock in the afternoon, a long while after we had passed the turnaround point. By this time, my optimism had disappeared, and I had shifted into what my field notes later recorded as "ordeal mode." My pace had become exceedingly slow and the experience of people passing me on the trail had become extremely irritating. My gaze had ceased almost entirely to fall upon the panoramas that were being brought into view, and my eyes now remained fixed on the stones a foot or two ahead of me on the trail. These were the elements of the landscape that now mattered, since they had acquired by this time an alarming capacity to become injurious. Jagged rock fragments and miniature boulders were now laid out in what could only charitably be called steps. It was impossible to repeat the actions of a previous movement to accomplish the one that followed. Each step had to be a new improvisation with its own coordination of aiming, pushing off, landing, and balancing. Despite my obsessive watchfulness, I miss-stepped frequently. There were no handrails to break a fall, and, as I felt how much they would have helped to save me from such a fate, my relation to the second-world Yosemite hiker construction began to disintegrate.

The park's developers and managers had intended that all hikers, weak or strong, would embody the innovation-demanding character of the trail. It produced—by physical shaping in this environment—a self-reliant, ever-ready visitor subject, a plucky, quick-witted persona who embodied some of the most admirable qualities of the Western American pioneer. This was the character that waited to be performatively reanimated at every moment of the trail's ascent. Even I, as I toiled slowly along—insofar as I was consumed with the need to keep working intelligently with each new set of stones—I, too, was internalizing the trail's second-world design, albeit in a manner more akin to that of the doomed Donner party than that of the fleet-footed idol of Yosemite trail culture, John Muir.[5]

Sometime after I had accepted that my calculations of the distance left to the top were seriously flawed, I put myself on fifteen-minute/five-minute climbing/resting intervals. This exertion-recuperation rhythm was not a structure the trail had been designed to afford and was not easily performed on these switchbacks. The trail was too narrow to allow hikers to pass by even if one rested standing at its edges. Nonetheless, the phrasing pattern constituted an attempt to cling to some form of conventional, culturally comprehensible structure. I trudged onward through an unstable maze of sharp edges, lopsided cracks and bulges, looking to my watch to measure my progress.

As the above details should make clear, ordeal mode was not at its core an intentional (second-world) phenomenon. It was the consequence of individual, physical limitations that had come into direct relation with the realities of the northern valley wall—its height, degree of exposure, and vertical incline. I had reached a degree of fatigue that caused me to lose interest in second-world constructions. Nothing seemed worth the effort I was currently expending.

The amount of stepping up required now felt "abusive," as my field notes later recorded. The route's continuously twisting, winding, ever-ascending form seemed "grotesque"—a hideous, monstrous, snake-like shape. The only meaning the top now held was as an end to this process once and for all. Yet, even while my hiking experience was being shaped by ordeal mode, I was also still acting as the pioneering figure, however fallibly. The two worlds were copresent during this stage of the hike, although the "real," first-world me was threatening at every moment to halt the safe progress of the second-world pioneer, bringing that tenacious, persevering character to its knees or worse.

Despite the bleak outlook, I did not for a moment consider turning back. It was still clear to me that I had what it took to finish the ascent. The rhythms of my actions and the regularity of my breathing patterns—these signs of self—still evidenced that the trail was not beyond my capability. No amount of dismal internal monologuing was going to persuade me otherwise. Unless the hazards presented by the steps got the better of me, a successful ascent was still at hand.

After a number of fifteen/five-minute intervals had passed, I began to resist the increasingly debilitating psychological patterning by silently reciting the Lord's Prayer. This was the only verbal composition I was able call up in memory at this point. Nothing in the trail's secular construction determined this choice of speech act. On the contrary, I had the vague sense that it was inappropriate. The decision was a choice determined by survival instinct as well as a much broader web of second-world symbolism that entangled the trail in various complex historical and cultural threads too convoluted and elaborate to trace out here in any detail. I recited slowly, synchronizing imagined enunciations with bodily actions, bringing myself into a single mentally and physically coordinated process of locomotion governed by the prayer's verbal rhythms. My stepping became vaguely reverential in character. When one foot touched the ground, and another shift of weight was successfully accomplished, the whole skeletomuscular system experienced a small surge of positive energy as it registered the understanding—literally the standing-under of the action just performed—that I was still uninjured. The silent verbal speech acts exerted a determining influence in this regard, preventing a breakdown of some sort. A second-world construct was still shaping, however shakily, the character of what was occurring in the first.

It was around this point on the trail, and when I was in this frame of mind, that the incident to which this account has been leading suddenly occurred. It was a new awareness of the valley itself. The awareness dawned as a concrete sense of the quality of the size of the valley—the whole of it, from top to bottom. I now registered, not visually, not conceptually, but energetically, through my moving actions, a vague but definite sense of the entirety of the way in which the valley rim connected to the valley floor. The valley began to make sense to me as if it could be gauged not in meters or yards or any other conventional formal measure, but in human actions and in the cumulative character of my own exertions—in the reserves that had been stored in and expended from my body's energy bank. I found myself somewhat surprisingly in receipt of an unnamed but still careful estimate of the will that it took, in general, and the movement that it took, in sum, to get up all of these infernal steps running all the way up to the valley rim and all the way back down to the valley floor.

As this awareness became more clear and vivid in my skeletomuscular system, I realized that I could now imaginatively comprehend, without any kind of conventional reckoning, the spatial scale of the valley. I had not gone looking for this understanding. I had neither expected nor hoped for it. It was not an intended goal of my hiking practice. Indeed, I had never imagined that anything remotely like this awareness was possible. I had never heard another hiker or visitor speak of it. It was an unconventional intelligence, formed in and through action, not from contemplation or discussion. It was the cumulative embodied result of standing-under the physical experiences of the hike—an articulate impression gained of it, as opposed to a conceptual representation of it. It emerged as an intelligible corporeal take on what the elevation gain of the valley's walls took out of me as a hiking self, and also of what I had put into hiking them over the years. It had not been born of constructionism's second world, although it very quickly had moved into it. On the contrary, it was of first-world origin, conceived and registered by the "real" me, what post-phenomenological theory would identify as the presubjective me, the living organism whose survival was on the line in this instance and who had learned by its own doing in this hazardous environment (Thrift 1996; 2000a; 2000b).

In terms of pragmaticist semiotic theory, what was happening in this incident can be described as an episode of semio-genesis (Chandler 2012), a process of "meaning-begetting" as Bruner has used the term (1986c, 13) or "meaning-making" as Sherry Ortner has characterized it (1999). It was a process that gave birth to a new sign, a sign of the landscape that was also a character of me. In the actions of hiking, an increasingly perceptually attuned connection, an increasingly close-knit engagement with the landscape's non-human-made, highly regular qualities of height and inherent gravitational force, evolved to a markedly different degree of familiarity in my physical being. As a consequence, there evolved an intelligibly different re-member-ing of the environment in my self. My actions in and of themselves consequently were now wise to the lay of the land as they had never been before.

The awareness I gained in this incident was not only an awareness of what and of how the landscape itself existed as an objective reality (an Other from a constructionist point of view). It was not simply an awareness of the unique quality of the valley's bigness. It was also a new awareness of myself. Moving along the trail now became an activity that was mind framing in a new way as far as the "I" that was participating subjectively in the process was concerned. I was no longer the "I" that had started out down at the trailhead below. "I" had moved into a new role—a relational figure whose character was forming and changing in connection with the landscape's living, changing processes, and with the laws governing them, human and nonhuman alike. "I" had inadvertently been revealed to me as a human measuring tape of sorts, a hiking inchworm that had unwittingly but nonetheless effectively recorded the spatial dimensions of the valley walls via its own movement performances.[6]

DESCENDING SWITCHBACK: SOCIAL CONSTRUCTIONISM

Changing one's mental awareness in an experienced, informatively growing way directly in relation to nonhuman features of a landscape, a process of objective learning, is inconceivable from social constructionist points of view. First-world realities simply cannot play such leading, determining roles in changing subjective awareness intelligently. Learning is not considered possible without some form of conceptual, discursive, or otherwise socioculturally fashioned second-world encoding that could recast the unknowable, meaningless real stuff of physical and material experience into socially and culturally recognizable terms. Yet, second-world encoding did not occur during the incident on the Yosemite Falls trail, and learning did take place—learning about how much the valley wall matter-ed to a first-world body and how much it had mattered over the years. Movements and their nonbuilt environment not only spoke louder than words in this learning process; they changed mental processes physically with a life-sustaining intelligence of their own patterning and design.

It might be argued from constructionist points of view that the interpretation of this incident overlooks the role of the Yosemite Falls trail itself as an intervening second-world structure. The trail was fundamental to the hiking process, and without it no relationship to the landscape's character could have been forged. As such, the trail served as a second-world mold into which the hiking experience of the valley's big-ness was cast. Certainly the trail did play a key role as mediating enabler in this incident, as similarly constructed trails had done in other incidents that factored into the learning process. Could I have learned what I did, and as I did, without the human-constructed trail? The answer would seem to be no. Therefore, didn't the trail's constructors and their second-world technologies and discourses determine my learning process?

A constructionist would answer yes, absolutely. However, my answer—and the answer forthcoming from a close reading of Bruner's ethnographic constructivism—would be no. My answer would be no, not because the trail was not a factor in the learning process, but because it did not play the determining role with regard to what it was that was learned in this instance. Moreover, it did not play the only role, as social constructionism argues; nor did it play a role that masked or blocked the first-world realities of the landscape and inserted arbitrarily constructed human-made signs in their place.

The trail's role was very far from deterministic in character, if it could be considered so at all. It did not, for example, determine the character of gravity as it acted on my bones and muscles, any more than it determined the character of gravity as it acted on its own physical materials. Likewise, the trail did not determine the size of the valley wall as that character affected my movements. Trail or no trail, these characters of the landscape were what they were. Moreover, the trail did not serve to separate me from any inherent qualities of the landscape by somehow obscuring or recoding their real nature. Rather, the trail related me to these qualities without displacement.

The semio-genetic process happening on and via the trail, in other words, was one of connecting copresences, mine and the valley's, not one of replacing one kind of character with another and dictating what could be understood from that substitution. To insist on the latter characterization of the trail's contribution to the experience gained would be to mistake its semiotic identity. The trail was, to be sure, acting as a sign of and in the landscape. However, it was doing so as a re-presenting sign, as an apparatus relating fluidly first-world realities, rather than recodifying or removing them. The trope of the arbitrarily crafted sign—the trope on which the symbolic determinism of social constructionism depends—fails as an apt characterization of the trail's semiotic character. It was a sign of conveyance, not opposition; of similarity, not difference; and of mediation, not removed representation.

To sum up, the awareness gained during the hike was the consequence of semiotic processes that were both human and nonhuman in design, both socioculturally and materially constructed. No social constructionist interpretation could account completely for the meaning-making character of this incident, given the range of semiotic agents contributing to its emergence. The air that was breathed, the gravity that was resisted, the sunlight and heat and the rock that supported and constituted the trail itself—all of these and other non-human-made features of the environment made their own contributions to the learning process, whose growth now informed and guided my practice and experience. In this integrated influencing, the valley walls and I were no longer absolutely or simply mutually Other. Now, at least in this semiotic reality, we had become living members of one ecological mind, partners in a shared truth concerning the valley's topological character that had been produced and shaped—in human-made words and in designs, to be sure, but also in bone and in muscle, in granite and in wind.

TRAIL'S END: BRUNER'S PRAGMATIC CONSTRUCTIONISM

This interpretive account, in what might be called its ecosemiotic realism, does not align with social constructionism in that it recognizes precisely that which constructionism rejects: objective significance or—in the terms employed above, first-world processes of learning. Indeed, I have formulated the narrative representing the Yosemite Falls trail hike specifically to focus on a moment in which I lived in and experienced the two worlds that constructionism claims are ever-separate as a single level playing field of emergent meaning-making—a field in which movements of semio-genesis were observed to flow in multiple directions. The account rejects the dualisms that social constructionism would reify: the opposition between what is real and what is meaningful, between what is experiential and what is significant or symbolic, between what is phenomenological and what is social, and between what is human and what is ecological. In this integrative regard, the interpretive approach taken differs markedly from social constructionism. However, it aligns with, not against, Bruner's constructivist ethnographic approach to the study of tourism.

Such a claim might seem puzzling, given the emphasis Bruner has placed on narrative forms of meaning-making and the central role his work assigns to acts of storytelling as the means by which tourists generate and construct their pre-, post-, and on-tour experiences as meaningful. It might be argued that no anthropologist has made a stronger methodological commitment to discursive practices such as interviewing and conversation than Bruner. As he stated in the introduction to his 2005 volume, *Culture on Tour*, "it is the story, the telling, that makes sense of it all, and the story is how people interpret their journey and their lives" (2005b, 20). The story has been Bruner's primary instrument of meaning-making in his observation of and participation with tourists, as well as in his interpretive analyses.

In the same 2005 introductory passage, Bruner goes so far as to accord exclusive agency to storytelling as the means by which sense-making occurs in tourist practices. He aligns his perspective with that of narrative theorists such as Jerome Bruner (2003), who have argued that narrative is necessary for any meaning to be made out of otherwise "inchoate experiences" (E. Bruner 2005b, 24). Edward Bruner characterizes somatic and affective forms of experience as "resistant" to verbal representation, and, in so doing, appears to reinforce the first/second world distinction of social constructionism (2005b, 25). Given the preeminent role that exclusively second-world symbolic practices of storytelling have played in his work, it would seem that his research supports even the strongest variants of social constructionism's symbolic determinism. "At the beginning and at the end," he has avowed, "the production of ethnography is framed by the dominant story" (1986b, 146).

Certainly, Edward Bruner has consistently relied upon discursive processes to pursue his study of tourism. However, despite the powerful place accorded to narrative in his work, Bruner has also and just as consistently rejected any move toward the discursive or linguistic determinism that social constructionist perspectives would insist upon. Bruner may have seen a disjuncture between what he has characterized as the inchoate world of experience and what he posits as the relatively meaningful structures of symbolic narrative, but he does not claim that the latter determine any understanding or awareness of the former. This is a crucial feature of his approach for the philosophical purposes at hand. Bruner's interpretive approach, in its very recognition of the resistance of somatic and affective experience to narrative formations, moves away from a deterministic view of the relation of constructionism's second world to its first. The resistance of the first world guarantees that no narrative structuring will ever negate its experiential contributions to human understanding.

This move away from linguistic determinism grants some measure of phenomenological (if not semiotic) autonomy to constructionism's first-world experiences. In addition, and far more important, it provides a philosophical basis for the claim that Bruner has made far more insistently about narrative meaning-making than any other: the claim that tourist storytelling is emergent, rather than merely reiterative, in its meaning-making character. Although experience may resist words for Bruner, it nonetheless can be imagined, felt, desired, and even remembered. In its memorability, in particular, it can serve as a source for the construction of individually creative narrative formations.

In his discussion of his experiences soaring above the African landscape in a hot air balloon, for example, Bruner writes, "These physical feelings are still powerfully inscribed in my memory of the trip and in my body, however resistant they may be to being put into words. I can feel them now, even as I write" (2005b, 25). Without such individual experience having a memorable life of its own, without it having its own "inscriptive" significance, as Bruner terms it, tourists would be far less able to do anything other than reiterate or contest in a reactive manner the discursive master narratives socially and culturally provided to them. However, the "mechanical" perspective on cultural transmission, as Bruner has characterized it, is not one he accepts (1986c, 12). Instead, he envisions a relationship between experience and narrative as one of "mutual interdependence" (2005b, 19).[7]

Bruner has also argued in favor of the view that the material attributes of tourist sites have agency in meaning-making processes. He characterizes them as being "alive" (2005b, 12, 25) in the same manner that I have described the Yosemite Falls trail and its material characters as animate and semio-genetic in this essay. The many possible responses of tourists that Bruner imagines evolve out of and are shaped by memorable non-discursively produced experiences and material conditions. They thus transcend or depart from preexisting narrative structures and move in new discursive directions. The relationship is not deterministic but rather "dialogical" (1986c, 6).

Bruner's ethnographies abound with examples of such emergent responses to tourism's master narratives, not only by tourists but by all those engaged in touristic encounters. In this regard, one thinks of the Maasai villagers Bruner observed living at Mayers Ranch in Kenya, who unexpectedly turned the tables on Bruner and Barbara Kirshenblatt-Gimblett during an interview in 1984 and started asking questions about the practice of female circumcision and polygamy in the American cultural context (Bruner and Kirshenblatt-Gimblett 1994, 465–66). Or of the example of the tourist at the Kichwa Tembo Tented Camp in Kenya who, to her complete surprise, received a marriage proposal for her daughter after having shown a photograph of the young woman to a Maasai guide (E. Bruner 2001, 896). One remembers the historically inappropriate flowerbed that suddenly appeared one morning in the New Salem reconstruction of Lincoln's childhood home (E. Bruner 1994, 401), and the waiter at the Kenyan Sundowner who spontaneously joined a traditional dance performance despite being completely inappropriately dressed (E. Bruner 2001, 896). Perhaps most memorable, however, and most representative of the phenomenological side of Bruner's approach, is the episode of the tourist journey by car through Tanzania, narrated in 2001, when the unexpected site of emaciated villagers along the way "broke through the bubble," in Bruner's words, leaving tourists with perceptions of Africa that their managers never meant for them to have, but which haunted them afterward, having been inscribed into their visual memories (2001, 901). Such episodes as these illustrate off-script departures from master narratives—the unexpected, surprising occurrences when tourism inspires creative, expressive responses.

Bruner's approach has thus foregrounded the emergent, creative propensities of tourists and tourism. Emergence, in this regard, must be recognized as a concept just as central to Bruner's perspective as is narrative. It is what motivates his ethnographic investigation of tourism—the tracking and mapping-out of emergent culture as it occurs in touristic practice (E. Bruner 2005b, 28). Whether the ethnography is situated in Bali or Kenya, in Ghana or Illinois, Bruner's goal has remained consistent: to witness and take part in what is newly being made out of tourism, not merely in what is inevitably always reshaped again and again. His ethnographic role has been one of being present to document what the cultural significance of emergent meaning-making might be. This definitive interest in semio-genesis, in the recognition that "culture is alive" (1986c, 12), experientially and symbolically, complicates Bruner's constructivist perspective, bringing it into alignment, in this specific regard, with relatively realistic pragmatic and post-phenomenological orientations.[8]

CAMPFIRE: BRUNER'S PERFORMANCE THEORY

There appear to be multiple reasons for the focus on the emergent, as opposed to the predetermined, character of meaning in Bruner's work. The influence of the realist tradition of American ethnography, in which Bruner was trained—a tradition that opposes the ontological and epistemological foundations of social constructionism—might be seen as providing some basic background support for his interest (2005b, 8). In addition to this formative influence, Bruner has pursued what he has termed a "radical" form of reflexive interpretive study (2005b, 9) that enables anthropologists "to merge with our subject matter" (1984a, 15) and to appreciate interpretive processes as experiences "in which the subject discovers himself" (1986c, 12).[9] This opens an epistemological door in his constructivism for the recognition of exactly the kinds of individually experienced processes that the Yosemite Falls trail incident illuminates. Influenced in this regard by the phenomenologically inclined hermeneutic philosopher Wilhelm Dilthey, Bruner's commitment to a reflexive, subjectively aware form of participant observation—one that he has characterized as moving "toward a more accurate representation of the world as it is" (1984a, 15)—acknowledges a wider range of creative experience than any deterministic social theory would allow.

Beyond these two influences lies a third, even more powerful influence that keeps Bruner's epistemological door not only open but swinging back and forth between phenomenology and hermeneutics. This third influence is the interdisciplinary influence of performance theory. In his attention to performance, both narrative and phenomenological, Bruner's constructivism becomes pragmatic in the Peircean sense—not only radically reflexive, but also radically inclusive, semiotically speaking. It is by engaging performance theory that Bruner is able straddle and negotiate between social constructionism's first and second worlds.

Bruner has argued that "the anthropology of performance is part of the anthropology of experience" (1986c, 7). His work evidences a sustained and guiding interest in tourism, narrative, and culture generally as critically performative phenomena. His interest in symbolic interactionism, his early engagement with Milton Singer's (1972) work on cultural performance, and, perhaps most significantly, his collaborations with performance studies scholar Kirshenblatt-Gimblett,[10] all appear to have produced in Bruner's approach a rich awareness of the ways in which symbolic processes occur as part of a larger performative world, always unfolding creatively in active relation to somatic, affective, and environmental factors, forces, actors, and stages.[11]

Bruner's sustained engagement with performance theory complicates his narrative approach. As he has consistently emphasized, his theory of narrative as it enters into ethnographic practice foregrounds its character as a "telling" phenomenon (1986b, 145, 153). His study of narrative is always a study of "performed texts" (1986c, 7)—that is, of stories told and retold—and his studies of culture are always studies not of texts in and of themselves, but of "the performance of text" and "re-performance" (1986c, 22) or retelling (1986b, 153):

> It is in the performance of an expression that we re-experience, re-live, re-create, re-tell, re-construct, and re-fashion our culture. The performance does not release a preexisting meaning that lies dormant in the text (Derrida 1974; Barthes 1974). Rather, the performance itself is constitutive. Meaning is always in the present, in the here-and-now, not in such past manifestations as historical origins or author intentions (1986c, 11).

The identification of the "re-"presentational capacity of narrative performance in this passage, and of the generative role that "re-ing" in and of itself can play in transforming lived experience into shared acts of meaning-making, aligns Bruner's perspective closely with Peirce's semiotics, as well as the approaches of pragmatic philosophers John Dewey and William James.[12] The performative framing compels the foregrounding of precisely that which is not determined by the master narratives or sociocultural scripts of touristic storytelling, but which instead relates to and draws meaningfully from actual life experience and what might be characterized as its unfolding living-ness or doing-ness. It is in this performative regard that Bruner's work can be appreciated as a basis for the analysis here presented, which is grounded as well in theories of cultural and touristic semio-genetic performance.

In sum, although Bruner's ethnographies, with their reliance upon narrative forms of meaning-making, don't always attempt to bridge the gap between social constructionism's first and second worlds, neither do they do anything to impose second-world constructs of meaning-making onto the memorable experiences originating in the first. On the contrary, they give as much recognition as possible to the ways in which first-world experiences matter in the lives of tourists and motivate their performances and their learning processes, narrative and otherwise. While Bruner's interpretive studies do not take what might be considered the next theoretical step that has been taken here—the step of showing how a bridging intelligence between the two worlds can emerge meaningfully in and move out from the experiential side—they nonetheless support and enable that step. They prepare the way for interpretive work that both illuminates and articulates what has previously been assumed to be inchoate: mindful movements of reflexive self–worlding, performative movements capable of prefiguring emergent re-presentational forms of meaning-making on and in the highly variable and rapidly evolving stages of tourism.

This ecosemiotic analysis of visitor activity in Yosemite National Park can thus be understood to belong to a theoretical trajectory that Bruner's constructivist research on the cultural performances of tourism initiated and has consistently sustained. What Bruner found to be beyond words in his own individual experience can be seen as the basis for creating new ways of relating to narrative and sense-making in the experience of others. The integrative key to understanding Bruner's approach as sympathetic to both pragmatic semiotic realism and social constructionism lies in the attention given, in both method and in theory, to performance as a process that connects even the most radically divergent modalities and productions of meaning-making.

NOTES

1. The National Park Service reports 3.9 million, 3.8 million, and over 4 million visitors in 2012, 2013, and 2014, respectively. See www.nps.gov/yose/learn/management/statistics.htm, accessed September 9, 2015.

2. See also www.nps.gov/yose/planyourvisit/yosemitefallstrail.htm (National Park Service website), accessed September 8, 2015.

3. As Pernecky observes, tourism literature uses the terms *constructionism* and *constructivism* inconsistently and sometimes interchangeably. I follow Pernecky, using *constructionism*, the broader term, despite the fact that Bruner used *constructivism* in relation to the epistemological issues framing his research.

4. Pernecky cites the work of D. R. Loseke in this regard (Loseke 1999).

5. The Donner Party was an ill-fated group of homesteaders who attempted to reach California by wagon train in 1846, but were delayed by a series of mishaps en route and then became trapped by an early blizzard in the Sierra Nevada mountains.

Forced to wait out the winter in very primitive conditions, more than half of the original group died of exposure and starvation.

6. The sign type identified here is a dicent indexical legisign, according to pragmaticist semiotics, and is similar to Peirce's example of a yardstick as an index (Peirce Edition Project 1998, 14, 461). This sign relation is evidenced by the fact that the hiking experience established a real connection between myself and the valley wall, and such a real connection is the key feature of an indexical sign. Because this particular hiking awareness was of a general character accumulating replications of its measuring significance, the indexical character, it qualifies as a legisign, and the "breaking news" conveyed as it reached consciousness qualifies it as dicent in Peirce's classification scheme. It could be seen as a kind of "natural proposition" extending from Frederik Stjernfelt's work on this sign classification (2014).

7. Bruner has consistently rejected any strong form of discursive determinism and identifies his perspective as "more phenomenological" than structuralist and deconstructive. He argues: "For me, no representation can fully exhaust the images, feelings, and meanings emergent from an event" (1984a, 7); and, "Master narratives provide a preexisting structure, but they are not determinative, nor can they encompass the many possible tourist responses" (2005b, 26).

8. With regard to pragmatism specifically, Bruner characterizes a common ground of sorts, writing, "These pragmatists focused on the processual; they did not take tradition or custom as a given, and they saw meaning as emergent rather than prior to events" (1986c, 14).

9. This quote is Bruner citing Wilhelm Dilthey (1976, 203).

10. For example, see Bruner and Kirshenblatt-Gimblett 1987; 1989; 1994. They also coauthored several conference papers during the same period.

11. See Bruner (2005b, 8, 9) and Di Giovine (this volume) for further discussion of influences.

12. See Bruner (1986c, 12–14) regarding his alignment with pragmatists Dewey and James.

Part IV

FURTHER AFIELD: NEW DEPARTURES

Chapter Seven

{Identity, Mobility, Embodiment}
"Being a Tourist in My (Own) Home"
Negotiating Identity between Tourism and Migration in Indonesia

Kathleen M. Adams

When I was an anthropology graduate student in the 1980s, my introduction to Edward Bruner's scholarship centered not on tourism, but on his contributions to the study of ethnicity, migration, and cultural performances in Indonesia. Although most readers of this volume will associate Bruner's name with his constructivist tourism research, when I was setting out to Indonesia to conduct fieldwork on tourism, art, and identity politics in the early 1980s, he had yet to turn his lens to tourism. However, the creative interplay between mobility, intergroup encounters, and identity was already a recurrent and fruitful theme in Bruner's writings, and it was this pretourism work that inspired my own approach to understanding and conceptualizing the politics of tourism in eastern Indonesia. This chapter showcases how Bruner's foundational work on these three analytical issues (mobility, intergroup encounters, and identity) is extended in my current research on Toraja migrant return visits to their ancestral homeland in the highlands of Sulawesi, Indonesia.

More specifically, this chapter examines the encounters, performances, and emotional experiences of different generations of Toraja migrants during their travels to the Toraja homeland. Toraja migrant returns are often inspired by desires to reunite with local kin or to partake in familial rituals, particularly funerals and ancestral house celebrations. Yet these visits generally also entail touristic activities and performances. Both touristic and kin encounters in the ancestral homeland tend to prompt reflections on self and the other. At the core of these reflections are narratives pertaining to identities, as well as struggles over contrasting meanings and understandings of various dimensions of Toraja identity (whether kin-based, spiritual, or newly emerging regional forms of identity).

Contrasting understandings of identity embraced by different parties involved in borderzone spaces is a recurrent motif in Bruner's tourism- and identity-oriented work (E. Bruner 1986b; 1991; 2005a). However, my aim here is to illustrate how a Bruner-inspired approach spotlighting the complex encounters, narratives, and performances pertaining to diaspora travelers and tourists can offer new insights into the processes of identity and kinship negotiation that are part and parcel of heritage tourism. A subsidiary theme in this chapter entails extending these identity-oriented explorations into the cyber era, examining how social media has given rise to new genre spaces for far-flung Torajas to engage in cyber-touristic identity-oriented explorations (K. Adams 2015a). Bruner's work with Indonesian migrants set the stage for questions that are at the heart of my current research.

TOYING WITH ETHNIC IDENTITY IN TOURISTIC SPACES: FROM MIGRANT BATAKS TO MIGRANT TORAJAS

As I outlined in an earlier chapter (this volume), Bruner's long-term research on Batak migrants in ethnically diverse Indonesian cities led him to chronicle the activities of Batak visitors at Jakarta's Beautiful Indonesia in Miniature Park. His broader aim was to tease out some of the identity-oriented paradoxes of flagship ethnic theme parks in several different countries (E. Bruner 2005d). Indonesia's ethnic theme park was born in an era of intense nation building and was designed to promote unity in diversity. (Indonesia is a multiethnic island nation with multiple religious groups, in which Muslim Javanese form the political and demographic majority population). As Bruner and others have speculated, an underlying aim of Indonesia's Beautiful Indonesia in Miniature Park was to deemphasize ethnicity and reframe subnational identities in terms of provincial borders and national orientations.[1] Thus one of the park's (then-) twenty-six provincial pavilions showcases Batak traditional architecture and cultural arts alongside the traditional architecture of other groups from the same province. However, Bruner's observations of Jakarta Batak families making recreational weekend visits to the park demonstrated that despite the park's design to emphasize provincial and national identity, these aspects of social identity are not salient to their visits. Rather, Jakarta-based migrant Toba Bataks celebrate and create their own versions of ethnic identity via their activities in the Batak area of the park, which stands in as a surrogate for the homeland. In short, these urban Bataks who grew up far from the ancestral homeland recreationally toy with ethnic identity in a touristic space in which they are "simultaneously the tourists and the toured" (2005d, 227).

Having visited the Batak display at the park and having made similar observations of Toraja visitors at the Toraja section of the South Sulawesi pavilion, I found the antiseptic feel of these ethnic displays striking. The

Batak domestic tourists Bruner observed explored their ethnic identity and heritage through a sanitized and distantiated version of their homeland. In this at-a-distance setting, idealized ethnic identities seldom have occasion to clash with the familial and clan identities given preeminence in the Batak homeland. Nor do urban Batak visitors to the park's Batak ethnic house encounter the same kind of disorienting assault of the senses that they might experience at a ritual in the homeland.

My own research with another Indonesian ethnic group follows different generations of Toraja migrants on their return homeland trips to address the question of what happens identity-wise when tourist/migrant-returnees vacation in the "unsanitized" homeland. How are at-a-distance idealized conceptions of identity and heritage enacted, reframed, or renarrativized? That is, how are ideas about self and other, home and not-home renegotiated in the context of these return visits? I turn now to discuss my ongoing work on these issues. Similar to Bruner's (1999) opening in his article on Batak migrants' relationship to their ancestral villages, which entails an account of his return to his Sumatran fieldwork site after a forty-year absence, I begin with a story about my own return, after an absence of five years, to the Indonesian village where I had worked.

TOURING ANCESTRY

In the summer of 2012, I returned to the Toraja highlands in Indonesia to attend the funeral of the Toraja woman in whose home I had lived on and off since the 1980s, when I began my long-term research in the region (see K. Adams 2006). This was the first time I had my husband and child (ten years old at the time) in tow, and I naively envisioned this visit more as a personal homage to my surrogate Toraja kin than as a foray down a new research path.

We arrived in the village ten days before the funeral, within hours of two other far-flung relatives of the deceased—a Toraja migrant and her Canadian-reared teenage daughter. Family members wasted no time in ushering us to the ancestral house to pay respects to my late fieldwork "mother," Sindo' Toding, who lay at rest in a lovingly tended coffin. Next we were escorted to the front of the family's ancestral house for a portrait with Sindo' Toding's partially finished, carved effigy (the latest in a long series of portraits with Sindo' Toding and family members taken on each of my returns). Finally, we all trudged down the bamboo-lined footpath to the cliffside village graves so the "American kin" visiting for the first time could meet and greet the effigies of other deceased relatives. The following day my adoptive fieldwork brother opened the family museum-cum-heirloom storehouse to offer my husband

and daughter what I later realized was an object-centered lesson in Toraja heritage. As I subsequently learned, the Toraja migrant and her Canadian-reared daughter's welcome mirrored ours.

Over the next few days, despite the pressing demands of preparing for the arrival of thousands of funeral guests, my surrogate Toraja kin orchestrated an unexpected series of activities for both the Canadian Toraja migrant family and my own family. One day, they whisked our families off on a mountain road trip, making stops for Toraja coffee and group portraits at scenic overlooks and graves. On another day, our hosts insisted our families join their entourage at a kinsman's funeral, so that the North American kin visiting for the first time "could experience a Toraja funeral." There, we perched on rice barn platforms and feasted while resident relatives introduced local and return-migrant kin and offered periodic tutorials on ritual protocol. The first-time visitors had all heard accounts of Toraja funerals from their Toraja parents, yet had not anticipated the novel sensory experiences: navigating the sticky, mud-coated ritual terrain while wearing sarongs; sitting a stone's throw from squealing pigs with foaming mouths and wriggling bound legs awaiting ritual slaughter; breathing air thick with cigarette smoke mingled with the musky stench of animal blood; and gingerly sampling (and sometimes savoring) unfamiliar foods (figure 7.1).

Figure 7.1. A guest reception procession navigates a muddied path bordered by pigs awaiting slaughter at a Toraja funeral.

Photo credit: Kathleen M. Adams.

Figure 7.2. A Toraja returning migrant-tourist photographing a funeral ritual.
Photo credit: Kathleen M. Adams.

Our languid afternoon at that funeral was periodically punctuated by the master of ceremony's crackling loudspeaker announcements, the Protestant pastor's sermon, bumbling attempts to sing from Toraja-language hymnals, and rearranging ourselves for flurries of group and selfie shots snapped by both local and returning migrant kin. Weeks later, back in Chicago, I discovered many of these selfie shots and portraits, along with others taken during Sindo' Toding's funeral, posted on the Facebook pages of migrant Toraja kin who had also returned for the ritual. One image in particular anchored itself in my mind and ultimately prompted a new course of research. It bore the caption "Being a tourist in my (own) home." Up until that point, my tourism-oriented research in the Toraja highlands had always focused on either foreign tourists or domestic tourists. However, sharing tours with returning diaspora Toraja kin and reading the visiting migrant's Facebook caption prompted the realization that I had missed an important cohort of people with complex relationships to the Toraja homeland that necessitated further attention.

In the Indonesian context, a rich, insightful anthropological literature on sociocultural issues pertaining to mobile persons has emerged; however, the overarching trend has been to spotlight either migrants or tourists, be they domestic or international.[2] To the best of my knowledge, Indonesian migrants returning to their ancestral homelands as tourists have yet to be the focus of extended study.[3] This is surprising, given that a variety of factors have prompted mushrooming numbers of Indonesian migrants to visit their ancestral homelands in recent decades, including the growth of budget airlines, the expansion of Indonesian destination marketing, and the promotion of ethnic homeland festivals on migrant-oriented social media pages.[4]

The Indonesian tourism ministry and industry approaches may be partially to blame for long eclipsing this cohort of mobile persons, as only two statistical categories exist in Indonesia: domestic tourists and international tourists. Likewise, Western academic biases surrounding our conceptual categories of tourists have contributed to a disproportionate emphasis on the ramifications of foreign tourism and while little attention is paid to migrants now returning to their ethnic homelands as tourists. As Tim Winter, Peggy Teo, and T. C. Chang observe, "the paradigm of tourism has in large part been constructed around an analysis of West-to-East, North-to-South encounters." In much of the scholarly literature, "our tourist has been silently conceived of as white (and male)" (2009, 5). Tim Winter (2009), Erik Cohen (2015), and others have underscored the need to decenter and decolonize the study of tourism in Asia by focusing on traveling Asians.

Since the late 1990s there has been some momentum in this direction within Indonesian studies. For instance, Shinji Yamashita (2003) has produced a nuanced study of female Japanese tourists who marry Balinese men they met while vacationing in Indonesia, and additional scholars (Pemberton 1994; K. Adams 1998; Hitchcock 1998) have turned their lenses toward various aspects of domestic tourism in Indonesia's ethnic theme parks. However, with the possible exception of Bruner's (2005d) aforementioned examination of domestic Batak tourists visiting Batak displays at Jakarta's Beautiful Indonesia in Miniature Park as a surrogate for their homeland, to the best of my knowledge, the activities of migrant-tourists visiting their Indonesian ancestral homelands are only beginning to attract scholarly attention.

The remainder of this chapter has three aims. First, I sketch some of the landscape of different generations of Toraja migrants' experiences on their return visits to their ancestral homeland. I am particularly interested in the ways in which these returning Torajas forge connections to their place of ancestry; that is, I examine how they engage with, react to, and emotionally process the peoples, places, and activities they encounter during the course of their visits. Ultimately, my broader research project is to achieve a better understanding of the ways in which migrants' visits to homelands may shift, challenge, or reaffirm sensibilities about their own identities.

In considering travel's relationship to aspects of migrants' identities, it is not my intention to advance a conception of ancestral homeland travel as a process that simply reaffirms returnees' irreducible, primordial identities. Rather, like Bruner and others, I embrace a constructivist conception of identity as always in the process of being formulated, challenged, affirmed, rethought, and remade. A return to a familial homeland can serve as a potent vehicle for questioning, reassessing, enacting, and articulating we–they relationships or contrasting sets of identities (see K. Adams 2016).

Nonetheless, it is valuable to underscore that in the special case of ancestral homeland travel, these we–they relationships are all the more complicated and involuted, as my Toraja migrant friend's Facebook comment about "being a tourist in my (own) home" suggests. As Shaul Kelner observes in his recent book on Jewish Americans' experiences on Israeli birthright tours:

> Homelands are gazed on not solely as a signifier of the Other, but also as a signifier of the Self. They hold out the promise of encountering difference within similarity, as well as similarity in spite of difference. In this dialectic resides homeland tourism's power as a locus for meditation on the meanings of diasporic identities. (Kelner 2010, 10)

The second theme I highlight in this chapter concerns the ironies and disconnects entailed in these homeland visits. The various generations of Toraja migrants I interviewed generally expressed longings, nostalgia, and pride in their ancestral culture, yet more often than not, their trips to pay homage to the homeland revealed either the superficiality of their cultural understandings, or disconnects between their perspectives and those of their locally based kin. These disconnects were particularly apparent with second- and third-generation Torajas visiting the ancestral homeland, reminiscent of the sharp contrast between African-American diaspora tourists' understandings of Ghana and those of local Ghanaians (E. Bruner 1996a). (As Bruner observed, African-American diaspora tourists tended to see Ghana as a stand-in for "Mother Africa" whereas local Ghanaians approached the African-American visitors as "white foreigners." Moreover, these visitors envisioned the Cape Coast castles and their dungeons as sacred ancestral sites of suffering, whereas Ghanaians tended to see these structures as more multifaceted sites with long and varied histories.) These differing perspectives engendered tensions and what Bruner termed "struggles over meanings" as Ghanaians sought to use their castles in ways that diaspora tourists deemed disrespectful. Although the Toraja migrants touring their homeland are only a generation or two removed, their visits can prompt similar disconnects, tensions, and struggles over meanings.

This brings me to the third theme, which pertains to the emotional terrain of Toraja migrant ancestral homeland visits. Bruner was ahead of most tourism scholars in recognizing the salience of the emotional realm in touristic experiences. This chapter draws inspiration from Bruner's attentiveness to emotions and meaning-making, and from more recent scholarship on the interplay between the realm of the senses and meaning-making in tourist spaces.[5] Crouch and Desforges (2003, 11) suggest that in attending to visitors' sensory engagements with locales we can gain unique insights into "issues such as subjectivity and inter-subjectivity, expressivity and emotion, memory and imagination." Writing about international tourists' travels to locales perceived to be ancestral homelands, Naomi Leite also stresses the importance of "sensing the past," observing that "sensory engagement is so effective because it is imaginatively experienced as being *the same* as that of one's own ancestors" (Leite 2005, 286–87, italics in original). She further notes that these tourists' sensory experiences in such destinations enable many to navigate across the eras of their displacement to an imagined past, thereby reframing their understandings of their own present-day identities and fostering a newfound connection to a "home"-place (2005, 297).

The Toraja returning migrant-tourists I write of here share much with the diaspora tourists Leite describes, in terms of the role of sensory experiences in shaping their understandings of their identities during and after these travels to the homeland. However, a key difference is that many of these Toraja homeland tourists are returning to villages in which they or their parents were reared. Their ties to these sites (and the kin that occupy them) are direct, yet the memories evoked by anticipated trips home are often selective, and they themselves have changed since their initial departures. Many return anticipating the pleasurable tastes and smells of longed-for homeland foods, the sustaining embraces of much-missed village kin, and the soothing breezes and memory-laden landscapes of youth. Others return dreading the financially and emotionally weighty burdens and entanglements of ancestral rituals. Yet, despite their direct ties to these destinations and their anticipated experiences, the emotional and sensory terrains they encounter on these homeland trips are not always as anticipated.

Here, I argue that the bodily and sensory experiences of Toraja migrants in their ancestral homeland visits entail cominglings of attraction and revulsion and foster potent feelings of ambivalence.[6] As some of the case studies in this chapter illustrate, ambivalence is a frequent yet underexplored theme in the narratives and comments of returning migrant-tourists.[7] I suggest that by examining emotionally charged, multisensual encounters with homeland kin, objects, foods, and animals we can gain more nuanced insights into how migrant visitors to their own or their parents' homelands feel their way to unique understandings concerning heritage and familial, ethnic, spiritual, and other identities.

THE TORAJA IN THE ZOOM LENS

Whereas Bruner's ethnographic research is heavily oriented toward the perspectives of visitors and other cultural interlopers (such as guides, anthropologists, and travel brokers), my research draws from insights culled from long-term ethnographic work amongst the Sa'dan Toraja, whose homeland is in the mountainous, predominantly rural interior of the island of Sulawesi. The Toraja are a largely Christian minority group in a predominantly Muslim nation. Over 750,000 Torajas reside in their homeland, and an equal number of first-, second-, and third-generation Torajas reside elsewhere, working or studying on other Indonesian islands or abroad. Toraja migration began in the 1970s, and over time remittances from migrants contributed to tremendous inflation in the cost and scale of rituals, as the more successful first-generation migrants funneled their familial status-enhancement efforts into cosponsoring elaborate, pageantry-filled funeral and house-consecration rituals in their homeland (K. Adams 2006; 2015a; Volkman 1985).

Funerals and other rituals not only draw first-generation migrants and their children back to the highlands, they also draw tourists: Toraja villages, rituals, and burial sites have been promoted for touristic consumption since the 1980s (K. Adams 2006). By the mid-1990s, roughly 50,000 international tourists and 200,000 domestic tourists were visiting annually (these figures include many returning migrant visitors with their families). Arrivals plummeted in the late 1990s and early 2000s due to a series of political, economic, and health crises in Indonesia. Thus, in 2006 government officials began developing initiatives aimed at attracting Toraja migrants and their families to return for homeland visits. Today, tourism to the highlands is recovering: in 2016 Northern Toraja Regency[8] alone received over 50,000 domestic tourists and over 60,000 foreign tourists, figures that reportedly included thousands of migrant Torajas returning for rituals, visits, and state-sponsored homecoming festivals (Kantor Staatistik Toraja Utara).

Returning to the ethnographic material, I begin with a discussion of the diversity of returning migrant-tourists and the range of factors that can color migrant experiences on their homeland visits. Then I showcase three themes: the ironies and disconnects entailed in migrants' visits, the emotional and bodily aspects of their visits, and the ways in which return visits shift, challenge, or reaffirm visitors' sensibilities about identities and heritage. Although artificially separated, these themes are, in actuality, intertwined.

DIVERSE MIGRANT EXPERIENCES

For some first- and second-generation migrants, poverty and geographical distance render return visits to the homeland challenging, if not impossible. Some second- and third-generation migrants have no direct contact with the homeland, though they may signal emotional connection to their heritage or ethnic pride by decorating their homes with Toraja textiles and carvings, or by joining Toraja-oriented Facebook groups. Others with closer kinship ties and enough income to cover the costs of bus trips and flights may make return trips once every few years, staying with local kin. Still others with more means—especially successful first-generation migrants—own grand second homes in the highlands and may return several times a year, sometimes with children and grandchildren in tow. In short, class and wealth intersect with migrants' experiences on their highland visits in various ways.

In terms of motives for ancestral homeland visits, all of the Toraja migrants I interviewed—wealthy or poor, first- or second-generation—underscored that they do not return to the highlands exclusively for tourism: rather, they return for rituals, to visit kin, to introduce their spouses and children to their cultural legacy, or to nourish longings for the homeland. While there, however, these first- and second-generation migrants combine those activities with visits to tourist sites. One construction worker living in East Kalimantan (Indonesian Borneo) told me, "We all feel homesickness. Like me, I've been away working (*merantau*) for thirty-five years. My wife is from Kalimantan, we've got five kids, and my whole family has already been to Toraja. Even my kids often return home to Toraja on holiday (*pulang berlibur*)." Another first-generation Toraja migrant working in Gabon's oil industry explained, "If I go home, I often visit tourist attractions there. I'm proud to be a son of Toraja soil. And, yeah, I also gulp in family (*men(j)eguk keluarga*), too."[9]

Comparing the typical experience of a wealthier Toraja migrant visiting the homeland with that of a migrant of ordinary means offers a glimpse into how migrants' financial status can color their physical or somatic experiences of the ancestral homeland. At present, when returning to see kin, attend rituals, and make pilgrimages to celebrated touristic sites, most Toraja first- and second-generation migrants of ordinary or more limited means reside with relatives, either in their rural homes or in specially constructed bamboo pavilions while attending funerals. There, they share food with their local kin, join in the rhythmic, melancholy *ma'badong* mortuary dances, and partake in playful banter while taking and posting selfies with village relatives, until the electronically amplified funeral music and chattering of the crowds die down enough for them to collapse, sardine-style, for a few hours of fitful sleep on mats or shared mattresses. After as long as ten days of sharing intimate space with local relations, despite physical discomforts, returning migrants told me of what one might describe as passing moments of *communitas* with local relatives and a sense of being in the fold of their ancestral culture. (See figure 7.3.)

In contrast, more affluent returning migrants and their children are accustomed to more comfortable urban households and less enchanted by the prospect of temporarily residing at a funeral site. They frequently express ambivalence about foregoing creature comforts to sleep alongside others on plaited mats in cramped funeral pavilions, trudging through mud spattered with sacrificial animal blood to queue for showers in semipublic settings. For some, especially second-generation migrants, these alien sensory experiences trigger powerful feelings of ambivalence and even revulsion. Many of the

Figure 7.3. Visiting migrant kin partake in a late-night funeral dance with their Toraja hosts.
Photo credit: Kathleen M. Adams.

more prosperous visiting migrants are prone to flee the crowds, acrid smells, and general discomforts of their kinspeople's funerals as soon as they can politely do so. As one wealthy Toraja migrant confided, after spending several hours each day at a rural funeral, he and his family "run away" to slumber in the cushioned comfort of lodges and hotels in the provincial capital. In short, the bodily experiences of these wealthier return migrants underscore a central irony: while they may be in their ancestral home, they are not at home.

THE WEIGHT OF HERITAGE: THE SOMATIC DIMENSION

Rante's story further illustrates the powerful role of the senses and bodily experiences in nourishing potent, visceral senses of connection to and disconnection from heritage. Rante is a second-generation Toraja migrant of aristocratic descent who was reared in Sulawesi's provincial capital. Most of the time when her parents returned to the highlands to visit kin, she opted to stay at home. As she tells me, "I grew up without knowing my culture. All I know is that Torajas have to sacrifice water buffalo when a relative dies, but I don't even know why." After graduating from the university, she met an American online, and they ultimately married. They settled in the United States and now have a young child. She once brought her husband and child to Sulawesi, and while there they visited kin and toured the Toraja highlands. She describes, with palpable euphoria, her experiences there of eating long-pined-for Toraja dishes. The tastes and textures of Toraja delicacies unobtainable elsewhere—fatty pig cooked in earthy black *pemerasan* (a tree pod seasoning) and fiery soup with gummy taro balls, greens, and chopped peanuts (*pogallu*)—transported her back in time to the nostalgic realm of childhood memories. It was through the sensations of taste and texture that she, quite literally, felt her heritage.

Yet this was not the case for her foreign-born child and husband. Her son wailed at the sight of ritual water buffalo sacrifices and sobbed incessantly when presented with buffalo meat thereafter. Still, she told me with a proud smile that her son relished pork, alluding to how Toraja sensibilities about ethnicity include a shared passion for pork—a powerful cultural identity marker in a Muslim country. Her American husband struggled with Toraja's spicy food and groaned about his incessant "Bali belly"[10] through much of the visit. In short, Rante and her family's contrasting bodily experiences pertaining to food offer a glimpse into the ways in which visceral experiences can nourish deep connections with place and ancestry for one generation while fueling revulsion in the next. Through this case we can also see how bodily experiences while on ancestral homeland visits can expose generational and cultural rifts within families.

For some second-generation Toraja visitors, a return to the homeland can prompt a different kind of sensory engagement with ancestors and heritage, an engagement that is not food-, people-, or animal-oriented, but more ephemeral. Elsi is a gregarious second-generation Toraja migrant in her mid-thirties. Elsi's parents, both Toraja aristocrats, left the homeland before she was born. They moved to Makassar and then to the United States to pursue higher degrees. When her parents returned to Makassar, Elsi stayed on for college and ultimately married a North American. When her father passed away earlier this year, she returned with her two young children to assist her mother with the funeral arrangements in the highlands. Between funeral activities, Elsi's brother and cousins whisked them off to see various tourist sites, all the while tutoring them in new, culturally prescribed ways of physically engaging with the landscape. On one occasion, they took Elsi and her children on a hike up Singki Hill overlooking North Toraja Regency's capital town of Rantepao. The aim was to visit a newer tourist site that had been erected in 2008. The site, which consists of an enormous cross and a large set of letters spelling out "Rantepao" (evocative of the Hollywood sign) is also an ancestral burial area. As Elsi told me,

> Walking up the stairs, we passed the ancient graves saying *tabe* (excuse me). *Tabe* is a kind of respect to the dead when passing their graves. Others believe that if you spit on the ground or just take a leak, you must say *tabe*. Even the sound of cars or motorcycles honking is [a way of saying] *tabe* [on the road]. On our walk back down, my cousin told us to say *tabe* many times because you don't want to be followed by ghosts (*Kunti Lanak*). My brother and cousin once went up with a group of tourists. The journey which took up fifteen to twenty minutes took them an hour or more. They felt something very heavy on their feet.

Elsi's local cousins shared many frightening Toraja ghost stories with them during their visit, and she found herself periodically having strange sensations. She recounted how they visited another new tourist site but abruptly stopped on their way up the freshly paved path as she suddenly experienced an eerie feeling while passing a thick cluster of greenery and bamboo. Her cousin then told her that a man had hung himself at that spot. Elsi also described how, the entire time she was in the highlands, people repeatedly told her they had seen her father's spirit, and she kept experiencing these strange feelings. Elsi's story illustrates how, for some second-generation Torajas, the highlands become a place where one can somatically experience the ancestors, where past and present merge through these sorts of embodied experiences.

THE WEIGHT OF HERITAGE: HOUSE-BASED FAMILIES, MIGRANTS, AND CLASHES OVER MEANING

Turning to the Bruner-inspired ethnographic strategy of unpacking clashes and disconnects between visitors and locals, I detour momentarily to offer background on the Toraja family structure. Despite seemingly familiar surface contours, Toraja family and social organization differ markedly from our own. Roxana Waterson (2003) has observed that Toraja social organization meshes with Claude Lévi-Strauss's notion of a "house society"—that is, a society in which kinship organization derives from named houses founded by ancestors. In house societies, ancestral houses are not only the key sites of ritual activity; they also own property and confer titles and duties. The Toraja ancestral house is intimately entwined with how people conceptualize familial identity.[11] In the broadest sense, Torajas call these origin houses *banua pa rapuan* (people who can trace their descent to a common ancestor constitute a *rapu*). The touristically celebrated, elaborately carved Toraja *tongkonan* is but one type of *banua pa rapuan*, associated with the nobility.

Each Toraja can claim membership in various family origin houses. Some of these ancestral houses are older, carry illustrious pedigrees, and are more prestigious than others. Less hallowed ancestral houses tend to be those that were more recently established, splintering off from the grander parent houses. One acquires one's house affiliations through both male and female parents and affines, and even more distant kin. These affiliations endure provided that one maintains ritual duties to the *tongkonan*, contributes funds to house consecrations, and repays prior generations' ritual debts by gifting water buffalo, which may cost tens of thousands of dollars, at funerals for house-based kin. Torajas' sense of house-based familial identity is so entrenched that Torajas encountering one another for the first time when far from the homeland frequently inquire as to each other's house affiliations (K. Adams 2015b).

Toraja sensibilities about obligations to family-origin houses also encompass protection of familial honor (*longko'*). As Waterson observes, "*longko*' can refer as much to feelings of shame in relation to other family members, as to feelings between families. . . . [These] feelings are equally elicited by a sense of one's failure to do the right thing, especially if one then incurs the scorn of other family members" (2009, 180). Not to sacrifice a magnificent water buffalo, not to return for a family funeral, not to repay an inherited funeral debt, even when one has migrated abroad, can shatter one's individual and familial honor (*longko'*). Participating in funerals, sacrificing livestock, and donating funds for ritual expenses help constitute family relations and ensure that familial honor remains intact.[12]

However, not all Toraja migrants fully grasp the nature and obligations of the Toraja house family. Accustomed to urban nuclear families, some second- and third-generation Toraja migrants have no knowledge of the ancestral house-based family system. For instance, when I asked Rante how she explained her Toraja family and their ritual obligations to her American husband, she said, "He just knows I have a *lot* of cousins, and that's really all I know, too." Other Toraja migrants understand house-based families but don't see the point of upholding *tongkonan* family prestige through engaging in costly rituals in the homeland. Thus some less wealthy migrants deliberately avoid returning to the highlands to avoid these financial obligations. They view house-based families as an economic noose no longer appropriate for today's times.

It is in this realm of contrasting understandings of the nature of the family that we find some of the most revealing and emotionally troubling clashes in understandings between second- and third-generation migrant-tourists and their local kin. The case of Tino Saroengallo illustrates how these clashes in sensibilities can become a source of anxiety, shame, ambivalence, and even anger.

Tino Saroengallo, a Jakarta-born Toraja filmmaker, describes planning his father's funeral in the family's ancestral village as one of the most emotionally haunting experiences of his life. As a youth, Tino's aristocratic Torajan father left the homeland on a scholarship to Leiden, where he met and married a Dutch woman. Together, they returned to Jakarta, where Tino and his sister were born and raised in an affluent home. Throughout Tino's youth, Tino's father held an important title in a celebrated *tongkonan*, and the family routinely returned to the highlands so his father could officiate at house-based rituals. On these trips, they stayed in their comfortable second home, and Tino and his sister were in vacation mode, visiting rituals and kin, enjoying mountain treks, photographing scenery, and making the rounds of tourist sites.

It was not until his father's 2003 death that Tino felt the emotional weight of his cultural heritage—a weight that made him physically dizzy (*pening*). As a freelance cinematographer with slim savings and a family to support, he had hoped to honor his father with a relatively simple highland funeral. For local kin, however, Tino's plans were shame-inducing, given his father's role in the ancestral house. They expected Tino to be fully invested in ensuring that the funeral was sufficiently grand to shield them all from the withering criticisms of rival house-based families. But, as an urban-reared, half-Dutch, second-generation migrant, Tino prioritized the economic needs of his own nuclear family and saw villagers' insistence that he stage an elaborate funeral as extravagant and backward. As Tino told me, on return visits to negotiate ritual details,[13] he experienced mounting resentment and anger over local house-family members' incessant demands for additional financial infusions. From Tino's perspective, he had already honored his father by agreeing to transport his body back from Jakarta for a highlands funeral. Ultimately Tino felt compelled to write a book about the experience in order to "exorcize" the negative emotions (Saroengallo 2010; see also K. Adams 2015a).

Years later, when I reencountered Tino in the highlands at his father's sister's funeral, he confided that the return visit had reanimated the earlier feelings of ambivalence and anxiety. Tino's emotionally haunting visits to his father's homeland offer insights into how migrants' homeland returns can engender profound emotional ambivalence and emotionally weighty clashes over fundamental understandings of aspects of one's heritage, such as the nature of the family and one's duties to kin.[14]

HERITAGE LITE: THE LOVELY DECEMBER FESTIVAL

One last story illustrates the growing economic, symbolic, and institutional role of Toraja migrant-tourists in the Toraja highlands, and highlights a new avenue for fostering feelings of connection to ancestral heritage. In late 2013, along with several other migrant relatives of my Toraja fieldwork family, I found myself Facebook tagged in an invitation to return for the "Lovely December Festival." This annual event, instituted by the Tana Toraja Regency government in 2006, when tourist visits had plummeted, was initially called the "Longing to Return Home to Toraja" Festival. Promoting the event on social media and in the news, the organizers hoped to convince migrant Torajas to make visits home in order to connect with family members who remained behind, as well as with fellow Torajas and the ancestral homeland. The ultimate aim was to boost hotel bookings and restaurant revenues during the tourism off-season. Homecoming parades, fireworks, contests, and other forms of entertainment, as well as hotel discounts, all added to the enticement strategy. While some locals grumbled about government expenditures for the outsider-oriented festival, it was an overwhelming success, and it is now an annual event that draws thousands of return migrant-tourists each December.

Many of the Toraja migrants I interviewed spoke excitedly about returning for the festival, Facebook-messaging me photos of themselves surrounded by festival fireworks or Christmas decorations. For many migrants contemplating a return to the homeland, the Lovely December Festival offers an opportunity to celebrate the less burdensome dimensions of ancestral heritage. As Lina, a thirty-five-year-old first-generation migrant married to an eastern Indonesian explained, "After my long journey, university in Yogya[karta], living in Jakarta and Bali and now Sengigi Islands, my feelings and thought patterns as a Toraja have changed. I don't want to oblige my kids to get tangled up in Toraja ritual customs, but the Lovely December Festival interests me. There's art, Toraja clothing, bamboo music, traditional dances." Lina's comments suggest part of the appeal of returning for the festival—the opportunity to connect to ancestral heritage, but not too deeply. The connections being forged are all aesthetic yet anesthetized, divorced from the realm of house-family duties and funeral debts. The festival's performances, activities, and displays share characteristics with the ethnic displays at Jakarta's Beautiful Indonesia in Miniature Park, described at the beginning of this chapter.

As Greg Acciaioli (1985) astutely observed of Suharto-era Indonesia over thirty years ago, in its confrontation with the diverse lifeways of its constituent peoples, the Indonesian state's approach was not to eradicate cultural diversity but to "emasculate" it, transforming complex and potentially troublesome cultural practices into spectacles. Writing of that era, he notes, "regional diversity is valued, honored . . . as long as it remains at the level of display, not belief, performance, not enactment" (1985, 161). For some visiting migrants, like Lina, it is in the sanitized, aestheticized performances of cultural heritage, stripped of elements that evoke emotional ambivalence, that more comfortable bodily and emotional connections to the homeland are found. In short, Lina and other migrants returning for the Lovely December Festival are consuming what I term "Heritage Lite." In the creative, sensory-infused yet emotionally safer touristic borderzone space of the festival, Lina has found a way to periodically reinhabit her ancestral culture.

CONCLUSION: HOMELAND VILLAGES AS IDENTITY SIRENS

Considering Bruner's triptych of research themes (mobility, migration, and identity) in light of the various returning migrant-tourist experiences recounted here, we see an array of ways of connecting, disconnecting, navigating, and renegotiating the meanings of identity, heritage, and home. However, some preliminary patterns emerge from the interviews and ethnographic research. First, returning migrant Toraja visitors (be they first- or second-generation) who stay in the homes and ritual pavilions of their local kin appear to experience passing moments of *communitas* with their local kin, seemingly forging deeper connections to their heritage than those who reside in the cocooning comfort of hotels. Second, returnees' visits to the ancestral homeland frequently entail emotional ambivalence—whether passing sensations or more profoundly haunting experiences. Yet as the Toraja cases suggest, the emotions, reactions, and sensibilities about ancestry, home, and belonging prompted by these returns are complex and far from uniform.

Bruner's observation prompted by his return to the Batak village where he had conducted research forty years earlier seems apt here: in our unbounded world, locality has "no monolithic meaning" (E. Bruner 1999, 475). For some Toraja tourist-migrants, the homeland signals a painful identity to be shed, for others it may be a place to somatically experience kin and ancestors, and for still others it is a space for reviving a romanticized version of familial or ethnic identity. The ancestral village may be all of these things—or none of them—at different moments in a single return voyage. However, for those migrants who choose to embark on a return visit, the ancestral village remains, for better or worse, a kind of identity siren, a beacon of possibility, danger, or promise. Many of the Toraja migrants who periodically vacation in the homeland might best be thought of as both home and not-home, or as living "in between" cultures, to use Yamashita's (2003, 99) phrase, forging their own unique spaces and shifting understandings of self and other. They warrant our attention as their lifestyle of navigation between cultures promises to continue in the twenty-first-century world.

NOTES

1. Others who have written on Indonesia's ethnic theme parks have made similar observations regarding President Suharto's motives for constructing the park (Pemberton 1994, 244; K. Adams 1998, 79).

2. For example, see the migration-oriented studies by Johan Lindquist (2008) and Rachel Silvey (2000; 2006), and the tourism studies by Michel Picard (1996), Jill Forshee (2001), and Stroma Cole (2008).

3. Shinji Yamashita's (2003) article on Japanese female tourists who marry Balinese men they met while vacationing in Indonesia is an exception to this, as is Bruner's (2005a) brief discussion. It was not until the early 2000s that scholars working in other parts of the world began to suggest that migration and tourism might be productively "understood as practices and discourses [that] are intertwined and mutually influence one another" (Intersections of Tourism and Migration Conference, Gdańsk, Poland, April 22–23, 2016. http://mobility-leisure.edu.pl/?p=572, accessed January 5, 2017).

4. Recently Indonesia's vice president invited the diaspora to help boost tourism in Malino, another mountainous region of South Sulawesi. See "Vice President Invites Diaspora to Boost Tourism in Malino, S. Sulawesi," *Jakarta Globe*, "http://jakartaglobe.id/news/vice-president-invites-diaspora-boost-tourism-malino-s-sulawesi/ (accessed September 1, 2017).

5. Although Finnish scholars Soile Veijola and Eeva Jokinen (1994) long ago decried the absences of the body in travel research, it was only a decade later that a growing number of tourism scholars began embracing the body in their studies.

6. I am grateful to Erik Cohen, who first prompted me to consider the dimension of attraction and revulsion in tourist experience.

7. At the time this chapter was written, Sabine Marschall's (2017) edited volume *Tourism and Memories of Home*, which addresses overlapping terrain via a wide range of case studies, had not yet been published.

8. The Regency of Tana Toraja was divided into two separate regencies in 2008.

9. Other Torajas reared away from the homeland feel obliged to visit the various highland Toraja tourist sites when returning with their migrant parents for rituals, as they know their non-Toraja workmates will inquire about these famed Toraja destinations. As one Toraja friend explained, "They don't want to appear dumb about their ancestors' homeland, since everyone in Indonesia has seen Toraja promoted on TV and online." For a fuller discussion, see K. Adams 2015a.

10. *Bali belly* is an expression some tourists use to describe indigestion.

11. Among the Toraja, these named house groups generally refer to physical structures that are periodically refurbished or rebuilt. However, in some instances, the names refer to structures that no longer exist in the physical world. For instance, one house group is named for an ancestral house that was destroyed in a fire long ago and was never rebuilt.

12. See K. Adams 2015a, where I also discuss the relationship between this concept and migrants' considerations about the pros and cons of returning to the homeland for ritual events.

13. Toraja funerals generally occur months or years after the biological death.

14. This chapter is dedicated to Tino Saroengallo, who passed away in July 2018, leaving a rich legacy of films and journalistic works. While battling cancer, Tino completed his final work, an achingly beautiful documentary film about the 2012 funeral of his Toraja aunt, Indo' Toding Sarongallo (*Sudah Selayaknya* [*A Proper Funeral*], Jakarta Media Syndication with SAE Institute, Geppetto Productions and Red Rooster Productions). Just months prior to his death, he traveled from Jakarta to the Toraja highlands to premier the film for his ancestral village kin, who were immensely proud of his accomplishments. This was to be his final gift to his rural relatives.

Chapter Eight

{The Self, Narrative, Borderzone} Beyond Dialogue

Hospitality and the Transformation of Self in Southwestern Madagascar

David Picard

The transformation of self is one of the darling topics of Edward Bruner's work on tourism theory. In a widely quoted article, Bruner (1991) observes the discrepancy between discourses about self and actual self-constructions in the field of tourism. He argues, for instance, that although the tourism industry and tourism culture at large promise tourists a transformational experience, they actually change little while on the tour. On the other hand, many destination societies, which the tourism industry and tourists often portray as ahistorical and culturally conservative, undergo substantial social and economic transformations when subjected to tourism development processes.

The aim of this chapter is to return to Bruner's engagements with the latter aspect of tourism and to explore his contribution to the study of creative social and cultural processes provoked by tourism and other external forces at the local scale. To illustrate the continued pertinence of Bruner's (1956a; 1961; 1991; 2005b; 2005c) approach to the formation of self as a dialogical process, I use data drawn from fieldwork I carried out in southwestern Madagascar between 2006 and 2011. At the same time, I push further Bruner's key concept of the borderzone and show how local actors not only become entangled in foreign narratives about local selves and culture, but may actively employ strategies of hospitality to engage, co-opt, and transform foreigners as a means of self-transformation and empowerment. I suggest that the focus on hospitality provides a conceptual tool that furthers and refines the investigation of the political, economic, and magical processes in and of the borderzone that is so central in Bruner's work.

NARRATIVES AND THE DIALOGICAL SELF

Deeply influenced by his training in symbolic interactionism at Chicago, Bruner emphasizes the centrality of the relation between narratives and experience for the formation and re-creation of social reality. He follows, in particular, the epistemic premises set out by Jerome Bruner (1994), for whom narratives are not primarily stories, but underlying master narratives that "make meaning, shape action, mold . . . behavior, serve to select which aspects of . . . culture will be displayed for visitors, [and] direct the construction of the infrastructure for foreigners" (E. Bruner 2005c, 3). Edward Bruner emphasizes that master narratives are deeply political in that they select what is important for people to experience, and what is not, and in that they supply people with a framework both to engage in a particular configuration of reality and later to think and talk about these experiences through conventionalized plots and story elements.

Focusing on tourists, Edward Bruner explains that they activate preexisting master narratives to engage in a role play, by means of which they enter the script proposed by the metanarratives associated with specific types of destinations, hence becoming actors of the tourist play. Pretour master narratives for tourists, according to Bruner, emerge through "tourist brochures, travel writing, guidebooks, television, film, scholarly accounts, anthropological writing, novels, journalism, and increasingly these days, on the internet" (2005c, 2). At different stages of the journey and in the post-tour context, the touristic role play experiences are mediated, once again, through narratives. In a way, tourism eventually allows tourists a personal appropriation of impersonal, public master narratives.

Through the implicit definition of positionality with regard to others, such master narratives and their mobilization in the tourism play allow tourists to reaffirm who they are as persons and how they relate to the wider world. And the experience of self and other is eventually articulated once again, through personalized public narratives of self and other (Turner and Bruner 1986). The link between person and group is maintained through the personal enactment of public master narratives, by means of which a collectively defined notion of self and the world are personalized. At the same time, Bruner suggests that selves remain mobile and in dialogue with a variety of narratives (2005b). They are never fixed in stone, but take shape through the interactions between personal sensibilities and experiences, and the structural confines of public narratives both preconfigure these experiences and help to add meaning to them. Building on the works of George Herbert Mead (1934), Bruner calls this the dialogical self.

Bruner (1991) stresses that in a similar fashion, people in destinations engage in dialogical narratives and role play, and reconfigure elements of self as scripted by the tourism industry at large. Bruner emphasizes that the power balance in these tourism borderzones is most often unequal, with storytelling rights attributed more often to tourists and their intermediaries, rather than to those whom the stories are about. He also contends that his concept of the borderzone is less politically charged or determined than concurrent concepts, especially the contact zone proposed by Mary Louise Pratt (1992). While Pratt suggests that local actors do actively co-opt and transform the narratives of dominant exogenous powers to engage in self-determined auto-ethnographic storytelling, she contends that the power balance is always unequal. Bruner observes that this is not necessarily the case, for most tourism borderzones are fluid and may shift over time; they are more akin to improvisational theater where both tourists and locals are actors (2005b). Echoing Renato Rosaldo's (1989) work, Bruner thus defines the borderzone as "a creative space, a site for the invention of culture on a massive scale. It is a festive, liberated zone, one that anthropology should investigate, not denigrate" (1996b, 159).

Just like tourists do not mechanically act out prescribed master narratives, people in destinations exercise a large degree of agency regarding the creation of meaning about roles they play within the frame set by initially external master narratives. This agency enables hosts to change or even pervert an externally attributed meaning or project into something different that makes sense from the perspectives of local actors and in contexts of local politics. In the early 1990s, Bruner's work on these counter processes constituted a major innovation in the field of tourism anthropology, which had previously been strongly dominated by structuralist approaches that posited tourism as a sacred journey (Turner 1973; Graburn 1977; 1983a) or a sociosemiotic field (MacCannell 1976; 2011). Bruner's actor-centered, dialogic perspective influenced a whole generation of younger researchers who adopted his approach: Michael Di Giovine (2009), Naomi Leite (2017), Tamás Régi (2013), Noel Salazar (2010), and Valerio Simoni (2016b)—and myself, in my work on hospitality and self-empowerment in Madagascar (D. Picard 2007; Picard and Moreira 2015; 2016), as I will develop in the following sections.

Early anthropologists conceptualized hospitality as a way to govern external life, especially to engage with mobile visitors who pass in and out. The main argument that I will develop in this work is that hospitality refines and deepens Bruner's approach to the transformations of self in the borderzone. I will show that in the Malagasy case, the symbolic and physical co-optation of external visitors becomes a principal means to form and empower local selves. The borderzone is not only a creative space in which cultures and selves are dialogically formulated and shaped, but also a space in which the presumed spiritual powers and substances of others are actively charmed, engaged, and appropriated in order to empower the self. We can observe here processes of social play and dialogue, and also of embodied co-optation and hospitality. Self not only is an abstract category of the human mind, as Mauss (1985 [1938]) put it, but becomes a social reality that materially and emotionally affects humans and their social worlds. For most of the Malagasy subjects who were part of this study, maintaining an ongoing privileged relation to specific foreigners—which can be humans but also spirits[1]—while guaranteeing these foreigners' welfare, luck, and health, becomes a condition for their own well-being (defined by the Malagasy notion of *hasina*).

APPROACH TO THE FIELD

When I arrived in the field in southwestern Madagascar in 2006, my initial research question was how and why coastal fishing populations would participate in marine conservation projects run by Western nature conservation organizations that had recently installed field sites in the country. This participation involved frequent and time-consuming stakeholder meetings with no immediate economic benefit to the fisheries. It was not immediately clear to me why the representatives of the fishing populations avidly participated and engaged with the foreigners. The Western workers of the conservation organization behaved, in a way, like tourists. They were highly mobile; most of them—especially a large number of volunteers—stayed for relatively short periods and usually never returned. Local populations, while periodically on the move during certain fishing or trading seasons (Astuti 1995; Cripps 2009), mostly remained in the area and hence witnessed the ongoing stream of different groups flowing in and through. In a slightly longer time frame of three to five years, local populations see the successive arrival and departure of international development projects that have been coming and going for the past couple of decades (Fauroux 2001; Pascal 2008). In this context, the borderzone has shaped, in its own right, a local metanarrative and social institution of hospitality towards these strangers that is central for the formation and empowerment of local selves, as a kind of political and magical power resource that local actors tap into to feed their social and political agendas (Chaudat 1993; J. Cole 2001; Evers 2002; Graeber 2007; Lambek 2013).

The key concept to study this context is hospitality. This is the practice of encompassing and co-opting outsiders in a controlled environment in an attempt both to neutralize their potential, unknown forces and to appropriate these forces to make them work for the hosts' own projects (Pitt-Rivers 1954; 1977; Candea and Da Col 2012). According Charlton Lewis and Charles Short (1879), the term *hospitality* is likely to come from a combination of the Latin noun *hostis* (meaning "stranger" or "enemy") and the Latin verb *petere* (meaning "exercise power over"). From the contraction of the two terms is derived the noun *hospitero*—literally, the "[one] exercising power over a stranger," but also "lord of strangers" or "the host"—and the verb *hospitor*, translated as "to be a guest" (Candea and Da Col 2012, who refer to Derrida 2000, who in turn refers to Benveniste 1969).

While this etymology offers an interesting sketch of the relational idioms linking host and guest in the Indo-European cultural tradition, the suggestion that the Latin noun *hostis* means both stranger and enemy is confusing given today's differentiation between the two words. The two terms are not equivalent: not every stranger is an enemy, and not every enemy is a stranger. Still, what both enemy and stranger seem to share is a symbolically and functionally equivalent quality of being a kind of exogenous force or power.

Many social scientists use Simmel's (1950) pioneering work on the stranger to understand the nature of this seemingly paradoxical meaning or value. Claude Lévi-Strauss (1970) and Joanna Overing Kaplan (1981) argue that among the Amazonians they studied, strangers—or others—are best dealt with by practices of cooking or cannibalistic consumption, which are efficient means of neutralizing the potential dangers of difference, but also, and maybe more importantly, of incorporating the presumed powers into one's own body and the body social (Rapport and Overing 2013). Hospitality hence supplies a concept that extends Bruner's approach to the dialogical self beyond the actual dialogue, and conceptually encompasses social practices of socially presumed substance transfer by means of which the powers of one social entity are believed to pass on to another.

SOA LINDA AND HER CATHOLIC SPIRIT

During my work in a fishing village south of Morombé on the southwest coast of Madagascar, I came across the case of girl whom I will call Soa Linda,[2] who was possessed by a powerful Catholic spirit. She told me that this spirit enjoyed singing and dancing and consuming the blood and body of Jesus as part of the Catholic liturgy. To keep the spirit vigorous and happy, Soa Linda regularly went to the church that Swiss Catholic missionaries had created and maintained in the village since the late 1960s. The head priest, with whom I talked on several occasions, remembered her as, in his words, one of the most devout Christians in her age group. She normally took part in the church ceremonies every morning before school. In recognition for this devotion, Soa Linda received a school fellowship for the church-run local private school, which eventually allowed her to pass secondary school and embark on a personal life trajectory that would not have been possible had the spirit been of a different kind.

Their identity is mutually constitutive. As long as Soa Linda was able to successfully play host to "her" Catholic spirit, the powers, skills, and protective abilities associated with this spirit would come into her possession. Over time Soa Linda and the spirit would become inseparable and turn into a common entity. While the narrative of possession suggests a possession of the human by the spirit (Lambek 2013), it is the human who seeks possession of the spirit, in a moral and material way. The narrative is actively appropriated into the lifeworld of Soa Linda and points to what Bruner (2005c, 2) calls an embodied experience of a narrative, one that exceeds the boundaries of the narrative of possession. The spirit or spirits in Soa Linda's possession and their respective vigorous, energetic conditions would define who she is, how and with whom she would interact, and, on the basis of these premises, which institutional frames and spaces would shape her life. This observation seems to corroborate a recent suggestion by Maurice Bloch (2011), for whom the making of self takes place at various scales or in layers, ranging from an intimate inner "minimal self," where—in my case—spirits become deeply formative of, and possibly indistinguishable from self, to formations of more fluid "narrative selves" including family, kin, and other identity markers.

THE POWER AND SKILLS OF FOREIGN SPIRITS

Soa Linda came to the village in the 1980s with her mother and maternal uncles, who had been contracted to work in a newly opened local hotel. Her mother, Helène, was a powerful *ombiasi* (a type of oracle or healer, and sometimes also a sorcerer), who brought with her a number of friendly spirits. When called upon, they took possession of her body and carried out various forms of diagnosis that identify problems of bad health or bad luck, to investigate suspicions of malicious attacks from other spirits, and they provided healing services or recipes to produce medicines and protective amulets (*fanafody*). Spirit hospitality hence indirectly transferred to Helène the technical skills of the spirits in her possession. The hospitable configuration was even more complex because many spirits with very specialized skills were not in her immediate possession, but had more or less amicable relations with spirits in her possession who acted as intermediaries. To contact these more specialized spirits, she therefore had to request mediation from the spirits in her possession, which often, but not always, worked.

The spirits in Helène's possession came to her one day, she explained to me, but they did not immediately reveal themselves—something that concurred with observations made in a number of other contexts in Madagascar (Sharp 1996; J. Cole 2001; Middleton 1999). In that phase, the spirits would typically not talk clearly and would mumble instead; a more powerful spirit was needed to force the mumbling spirits to out themselves. During a specific ceremony, another *ombiasi* would be called on whose spirits would threaten Helène's—with a spear, a knife, or bare fists—to force them to come out. This they eventually did. Once her daughter, Soa Linda, became a teenager, she was, in turn, possessed by first one, and later two more spirits. As in Helène's case, the spirits did not initially reveal themselves, and this time Helène called upon her own spirits to insist that those that had come upon her daughter reveal themselves. In a semipublic ceremony, a loud struggle unfolded (captured on camera by a local gendarme), in the course of which Soa Linda's spirits eventually capitulated and started to speak clearly. In both cases, it was the power and skills of the spirits in the possession of humans that allowed them to force another spirit into submission.

The observation of such a socially held general logic was by no means restricted to Malagasy people. Caterine and her daughter seemed to believe that the Swiss missionaries had brought with them their own spirits, namely the powerful Holy Ghost, who allowed them to successfully establish a church, build schools, and become important sociopolitical actors at a microlocal scale. In the late 2000s, an Italian medical NGO built a hospital facility on the borders of the village, and many people went there to seek diagnosis and treatment for various types of diseases and disorders. The leading Italian medic, before performing a diagnosis or surgical operation, was seen praying. He spread his arms wide open to the sky, murmuring a prayer, not unlike the way in which a spirit medium or *ombiasi* would call upon a spirit. Just like the Swiss missionaries, the medic was considered an *ombiasi* working with the help of powerful spirits. For many people it was the spirit who came upon him who performed the diagnosis and operation. When interviewed, one of the Italian medics who was religious and the Swiss head missionary agreed with such an understanding of the exogenous nature of medical and healing skills, provided by holy forces.

The Italian medics, like the Catholic missionaries, were able to perform many, but not all, services that were asked of them. A critical case I followed was that of Viviane, who had diabetes and had developed gangrene in one of her feet. The Italian medics told her to come to the hospital every day to receive an injection of insulin, which would attenuate the symptoms. Yet after only a few days she gave up the treatment because the Italians (and their spirits) seemed unable to heal her condition, she explained. She moved on to see other *ombiasi*, while her gangrene progressively worsened. Viviane later died of what appeared to be blood poisoning.

In the context of this study, practices of diagnosis and healing not only concerned pathological conditions in the Western biomedical sense, but equally conditions of bad luck or the perceived loss of power (loss of *hasina*)—for example, in the exercise of fishing activities. When the nets remained empty, the fishermen usually considered this a result of a malicious attack by jealous neighbors or family members. In the cases I observed, they would consult an *ombiasi*, who asked a specialist oracle spirit to diagnose the source of the problem—usually a malicious spirit—and to chase the latter away. Such a context emerged during the early 2000s when many fishers observed a dramatic degradation of the fisheries (essentially due to ecological changes in the coastal environment caused by the consequences of global warming, coral reef destruction, and the 2004 tsunami). The spirits the fishers called on to solve these environmental issues proved relatively helpless. The fisheries continued to deteriorate, thus challenging the powers of the established families and their spirits.

In this context, a British marine conservation NGO that had started work in the area in the early 2000s suggested the development of no-take zones, where fishing would be temporarily or permanently prohibited so juvenile fish and corals could reproduce. A demonstration project, based on a temporary no-fish zone for octopuses, had proved highly successful (Lill 2011). As if by a miracle the number of octopuses had disproportionally increased in the temporarily closed-off area. Observing the apparent power of (the spirits of) the British scientists to predict or even act on the future, many fishers desired to have their own marine park devices installed near the sites they used for fishing. However, the scientific rationale underlying the idea of a marine park was mostly lost. In some cases, fishers argued for the installation of no-fish zones in areas that were not fished anyway (for example, in areas that were too far away, or in deep water), with the idea that the device itself would provoke increased fish reproduction. For many fishers, the very device of the no-fish zone came to be considered a tool to reinvigorate the abundance of fish and the income generated by the fisheries sector.

Eventually, the fishers realized that the devices installed by the British NGO did not help much to increase the number of fish. In the late 2000s many started to abandon their collaboration with the marine conservation project—as if their spirits had proved insufficiently powerful to resolve the overall problem, like the Italian medics who could not cure diabetes.

COMPETING OVER THE RIGHT TO OFFER HOSPITALITY

In the borderzones of spirit hospitality, different social actors, including spirits themselves, compete over variously skilled and powerful spirits, and over the rights to access them. In this context, people and family groups are always at risk of losing some of their powers, or simply disappearing as selves, once their spirit-guests lose power or become insufficiently skilled, or are lured away by other players. This risk seems to increase in contexts of rapid ecological, economic, or social change, when existing livelihoods disappear and new opportunities arise.

One of the most prominent emerging political leaders in the village was Jean. Originally from a neighboring village, he had married into a branch of the local royal family and thus would normally have remained socially marginal. Yet Jean's social and political ascension was sourced through direct privileged contact with the most powerful foreigners and foreign[3] institutions to emerge locally. As a young man he was a soldier in the French colonial army, and he later worked for foreign cash crop collectors. After the communist revolution, throughout the 1970s and 1980s, he was the local representative of the national communist government. Following the political liberalization of Madagascar in the 1990s, he became the chief of the village elders, playing a crucial role, for instance, in the privatization of land that followed the 2005 land reform law (Picard and Moreira 2015).

In the 2000s, when the British marine conservation NGO and a number of newly installed foreign hotel developers started to work in the area, they systematically called on Jean to perform ceremonies to inaugurate conservation or building projects. He was the man to get ahold of when "you want things done," as one of these foreign operators put it to me. As part of the ceremonies Jean slaughtered cattle paid for by the foreigners, and enacted the role of seigneur in distributing meat among his family and clientele, thus keeping everyone's spirits happy. Jean responded to the expectations of the British NGO toward their local hosts to improve his economic and social position. What Bruner (1996b, 158–59) explains for the borderzone between tourist and locals holds true in this context, too; Jean understood the expectations and narratives of British NGO's and engaged in strategic performances of hospitality to enhance his political power and economic wealth.

The current chief of the local royal family, Lahatra, observed Jean's progressive empowerment without enthusiasm. He repeatedly explained to me that he was not happy about the situation, insisting that foreigners should work with him, not Jean, whom he called an imposter. Lahatra claimed that Jean consulted spirits that were his because they were inherited from Kipake, his father.

During the opening ceremony of the above-mentioned NGO-managed octopus reserve in 2006, a young girl from the village drowned in shallow water. Lahatra claimed that because he, as the rightful person to contact the spirits, had not been invited to perform the ceremony, the spirits, in turn, did not know about the opening, and hence did not provide their protection; therefore the girl died. He asserted that the spirits simply ignored the ceremony performed by Jean, or possibly even actively punished Jean via the death of the girl. In the days that followed, controversy over interpretations of the drowning developed, involving different *ombiasi* and their respective spirits. The situation was not resolved: Jean's side claimed that "God" had decided to take the girl back, while Lahatra's continued to claim that the spirits had not effectively been consulted and hence did not provide protection.

The case demonstrates the nature of ongoing competition over access to spirits and the powers they can provide. The techniques used in contestation include rhetoric, witchcraft, the smart tactical deployment of alliances, and strategic positioning in the contact zone with foreigners. In some cases it also depends simply on luck (which is mostly related to witchcraft): being in the right time and position to create a privileged relationship with foreigners and their spirits.

There is further rhetorical struggle when people try to convince others, including immediate family, of the power of their own spirits, and the lies or pretentions of others. Jean and Lahatra, for example, systematically questioned the very basis of each other's art and existence. The situation created a climate of constant suspicion and fear among the main actors and their families, which rarely led to open collision, but was transposed to the more hidden realms of witchcraft practice. For most people I talked to, one of their main daily concerns was to protect themselves and their families from spirit attacks and bad luck, once again by nurturing good relations with one's own spirits and attempting to encompass and domesticate other powerful ones, especially those of foreigners.

HOSPITALITY AND SELF-TRANSFORMATION

Even if Soa Linda's participation in the Christian liturgy was initially inscribed in a social logic of spirit hospitality and possession, through its continued and ongoing repetitions it had a disciplining and transformative effect. Over time it generated a specific form of social participation and belonging, and also shaped a specific worldview influenced at least partly by the moral rules imposed by the Catholic Church with regard to sex, drugs, and family relations. Practicing contact with her Catholic spirit hence deeply reshaped the self-perception and life contexts of Soa Linda's self, giving her access to new forms of sociability and also to different professional careers.

What this case indicates is that hospitality may well be geared at foreign visitors, but it will eventually transform the lifeworlds of hosts themselves. The concept of master narratives and of their effects on the lifeworlds of tourists and vice versa, employed by Bruner (2005a; 2005c), can be extended to local hosts too. Systematic analysis of the family backgrounds of all the children from the village who successfully passed through secondary education over the past fifteen years shows that all the most financially successful fishing families adhered to the Christian liturgy (Picard and Moreira 2016). There may be two general explanations. First, non-Christian fishers may have a stronger tendency to seek luck and protection through private and hidden relations with an *ombiasi* and their spirits, while Christian fishers may tend to seek the same luck and protection through semipublic relations with the Catholic *ombiasi* (the priest). In most cases, both forms of relationship continued concurrently. The moral doctrine of the church is relatively transparent, transcribed in the form of basic rules—regarding marriage, sex, alcohol use, humility, charitable acts, having a "clean spirit," and so forth—that emphasize individual behavior and social community, potentially generating a social, rather than individual, basis for ongoing financial success. Such rules of behavior and attention to the greater good are not part of the relationship with the (non-Christian) spirits contacted by most *ombiasi*: one consults them, pays, and gets whatever the spirit is willing to give—services of healing, but also of harming others—without much further social control.

A second explanation may lie in the potentially higher financial and economic efficiency of the kinship model propagated by the Catholic Church. In the customary family model, children can be adopted into the family of the biological father or the husband of the mother, or, on rare occasions, someone else's family (Randrianantenaina 2009). The new father then sends a considerable amount of money, alcohol, and cattle to the family of the mother, to appease their spirits and allow the transfer of family belonging. About half of the children researchers met at the village school had gone through such an adoption rite (*Sorognanake Vezo*); the other half remained with their mothers, in matrilocal households.

It is plausible to think that in a household shaped by the model of the Catholic nuclear family, the surplus from economic activities stays to a larger extent within the nuclear household where it may be invested in education, the betterment of houses, and the expansion of economic activities. My observations in the communities seem to confirm this hypothesis: those more actively participating in Catholic liturgy were generally financially richer, had more children at school, and had better houses. Many had moved their livelihood bases from artisan fishery activities to fish collection and trading. The non-Catholics, on the contrary, seemed to spend far more money on sorcerers, alcohol, and the maintenance of multiple love relationships. They also seemed more prone to move into higher-risk activities like offshore shark fishing, sea cucumber fishing based on apnea diving, and seasonal migrations to the far north.

CONCLUSION

Bruner approaches self-transformation in the field of tourism as a dialectical process whereby socially framed experiences of self and the world are articulated through the narrative expressions of such experiences. The central analytical focus is on the formation, mobilization, and dialogical transformation of master narratives of self and other that help tourists and hosts to interact with each other and to make meaning of their interaction experience. The innovation this approach brings to the field of tourism studies allows researchers to investigate the political dimension of tourism communication, whereby words, objects, symbols, and story plots are mobilized by different social actors, to articulate social hierarchies, inclusions, and exclusions.

Bruner's actor-centered approach can be usefully extended to the study of hospitable relations, as in this case with various types of spirits. Yet dialogue goes beyond social relations of exchanging words, signs, and gestures, and extends into the field of physical and spiritual co-optation and believed transmission of qualities and substances. A self without spirits would be stripped of identity, power, and *hasina*. From that perspective, everyday life is structured not only by the ongoing practices not only of dialoguing with spirits but, more importantly, of nurturing one's spirits' health and well-being as a means to enjoy their protection, knowledge, skills, and power. At the same time, these processes take place in a contested space in which effective possession of a spirit is never a given, but is challenged by others' attempts to seduce and appropriate one's spirits and lure them away. These processes are subject to a large degree of fluidity, self-innovative capacity, and contextual accidents. The power of self can increase in intensity but can also falter when spirits gain or lose power, or when their relations to humans change.

When Western foreigners—Swiss missionaries, Italian medics, British marine conservationists, and Western ecotourists—appear within this borderzone, they are considered primarily in terms of the powerful spirits they bring with them, simultaneously feared by local hosts for their unpredictability, and desired for their potential to increase political power. The foreigners are subjected to various forms of ceremonial hospitality, in the course of which they are incorporated into the body social and shown, if not offered to, the host's spirit pantheon. I have observed that the attempts to encompass the spirits of these foreigners eventually fail when the foreigners act against the expectations related to the hospitality they are being offered and to the wider cosmology of spirit beliefs in which it is embedded. The relation between agency and the acting out of norms and rules is never fully predictable; social reality remains a constant process.

NOTES

1. While tourism and touristic hospitality provide an overall analytical framework for this work, the Western semantics of the words *tourist* and *tourism* are of relatively limited importance because Malagasy hosts do not use these to distinguish between different types of foreign visitors—whether Western or national tourists. Instead the emic notions of *vazaha* (external foreigner) and *vahiny* (national foreigner) are the relevant concepts Malagasy, and therefore I, use.
2. This and all other names mentioned in this chapter are pseudonyms.
3. Foreign in the sense of extralocal, hence including other Gasy (*vahiny*) as well as non-Gasy (*vazaha*) people, governance agents, and institutions.

Chapter Nine

{Contested Sites, Identity, Stories} "Ideologies at War" at Chichén Itzá

An Ethnography of a Tourism Destination

Quetzil E. Castañeda

This chapter is an ethnography of Chichén Itzá, Yucatán, México, as a storied, performed, and contested space of tourism. I discuss, on the one hand, Edward Bruner's dialogical approach, which prioritizes dialogue not simply in a literal sense, but also in terms of what he refers to as dialogical narration and processes (Bruner and Gorfain 1984). On the other hand, I present the concept of "total social fact," borrowed from Marcel Mauss, and the analytical method of "extra-vagant anthropology," developed by James A. Boon (1999), as a methodological framework that is related to but different from Bruner's approach. In the main body of the chapter, I use a storytelling style of ethnography to describe and analyze an event as a total social fact by interweaving stories, events, texts, and representations that crisscross in and around this event.

This chapter, and my work in general, is inspired by Boon's extra-vagant anthropology (see Boon 1982; 1999; Whitmarish and Dent 2017; Castañeda 2017); my approach both converges with and diverges from Bruner's tourism ethnographies. The word *vagant*, as well as *vagrant*, are derived from the Latin root *vagari*, meaning "to wander, to roam," and "homelessness," respectively; the meanings of these words, including *extra*vagance—which points to both the extraordinary and *extra*-ordinary as subject matter of anthropology—migrate into each other. Boon's use of this assemblage of meanings points to the concept of the travelogue in the senses of both traveling logic and story or narrative of travel. Bruner (1986b) prominently pointed out in the 1980s what is today a truism: ethnography is narrative, a form of storytelling. Extending this assumption with Boon's notion of extra-vagance as well as Michel de Certeau's (1984, 91–130) idea that stories are travel practices and performances of stories, this article begins with the understanding that ethnographic analysis is travelogue in several different senses and ways. Extra-vagant anthropology as a methodological guide does not therefore proceed in a straight line to connect self-evident facts; rather this concept of ethnographic storytelling is a journey with detours, side trips, wandering, curving, circling back, and crisscrossing of narratives, texts, events, symbols, and experiences. Ethnographic understandings and explanations are created in and through this crisscrossing of social facts.

Significantly Boon and Bruner share some intellectual sources and traditions. Both were active protagonists in creating new kinds of interpretivist and constructionist methodologies that emerged in US cultural anthropology in the 1980s. Their ethnographic approaches both put stories and narratives up front as objects of study and as methods of analysis. Although both are concerned with the dynamics between performance, event, stories, text, and representations, they each place different emphasis and priority on these objects of study, which in turn leads to divergence in style, goals, and modes of ethnographic storytelling. Thus although both have participated in and advanced a constructionist tradition of anthropology, there are some important divergences that led Bruner to explicitly dismiss as inadequate text- and representation-based versus dialogical and performative approaches (e.g., Bruner and Gorfain 1984, 56–58). This chapter juxtaposes Boon and Bruner to perform a text-oriented ethnographic storytelling and analysis of Chichén Itzá as a contested site. In the conclusion I summarize differences between Bruner's methodology of "processual dialogism" (Bruner and Gorfain 1984, 56–60) and this Boon-inspired travelogical ethnography of and extra-vagant approach to tourism.

BRUNER'S DIALOGICAL APPROACH TO CONTESTED SITES

Anthropologists have always inherently studied contested sites, but not necessarily as overt objects of study. In the 1980s place and conflicts in and about such spaces became explicit problems for the social sciences in new ways, with new theories and concepts. The first generation, or wave, of anthropologists studying tourism, from the early 1970s to early 1990s, were definitely concerned with tourism destinations, but not in an analytical or methodological, much less theoretical, sense of space and sites; rather, destinations were mostly conceived generically as cultural contexts, undefined communities or localized places in which to study questions of host and guest roles, interactions and consequences for cultural communities, or tourist activities in terms of ritual, cultural bonding, consumption, and identity making.

Similarly the issue of conflicts over the meaning of these destinations was most often of two specific, restricted types. Contestation of meaning might revolve around whether something was or was not authentic for different kinds of locals and for different kinds of outsiders (tourists and ethnographers who interpreted a phenomenon as authentic or as some form of constructed authenticity). Or the contestation of meaning revolved around whether the effects of tourism—or impact—were good or bad or took this or that shape. Just as often, these two ways of studying conflicts converged as a matter of the meaning of place that was held by a homogenized insider host versus a similarly idealized outsider guest (or tourist), and an equally outsider ethnographer interpreting the (singular, unified) meaning. Greenwood (1977) and Stanton (1977) conducted two (quite opposite) types of studies that are exemplary of these earlier approaches that Bruner's work supersedes. Bruner broke from this paradigm by understanding that meaning is never singular or unified but is inherently multiple and therefore contested in its expression and communication in words, stories, and performances.

This assumption led Bruner to find conflicts over the meaning of places between multiple different actors, both within and across groups of locals, visitors, experts, and institutions (see Bruner and Gorfain 1984; E. Bruner 1993b; 1994; 1996a). Further, his dialogical approach shifted his focus from overt sociopolitical and economic conflicts to a new—for the anthropology of tourism—conceptualization of conflict as an object of study and method of its analysis. These conflicts manifest not so much in confrontations, physical or institutional acts of aggression, and antagonisms between people, but rather in the friction, tensions, strains, disjunctions, disputes, and disagreements between competing (or alternative) understandings, values, meanings, and choices as expressed in stories, dialogue, and performance. In particular, his attention focused specifically on the conflicts of meaning among and between different kinds of official, expert, and lay interpretations of places based in sociocultural identities (see Bruner and Gorfain 1984; E. Bruner 1993b; 1994; 1996a). Social scientists use the umbrella word *contestation* to sort together all these various forms of low intensity friction and conflicts over meaning. Among the possible different theoretical approaches to studying meaning that are available to anthropologists, Bruner explicitly invoked Mikhail Bakhtin's dialogical theory of speech communication. Thus Bruner envisioned and described this second type of contestation as the target of his study, which he isolated for analysis in phrases such as "ideologies at war," "dialogical narration," "dynamic chorus," "polyphonic stories," "dissident voices," and "dialogical interplay" (Bruner and Gorfain 1984, 56–58; E. Bruner 1993b, 14, 23).

In this chapter I also rely on a dialogical method of identifying objects of study and of analysis, one that is derived from Boon and from Vincent Crapanzano (1991; 1992). Boon's method is dialogical in the sense that he does close analysis of "what is said" (i.e., stories, texts, practices) and the way it is told, performed, or enacted in order to reveal the similarities and differences between the meanings expressed by different actors. Crapanzano's analytical method reveals conflicts, contradictions, and ambiguities of meaning at the level of words, word choice, dialogues, and discourses. Both of these orientations are expressed in my analysis not of the contested meanings *of* a site, as Bruner has studied (at Elmina Castle, Lincoln's New Salem, Mayers Ranch, and Masada), but of the contestations of meanings, politics, and practices that are enacted *in* a site of tourism. In this way I change the ethnographic object of analysis from conflicting interpretations of the destination to the hidden and muted, if not invisible and silent, conflicts within words, discourses, ideologies, practices, events, and ownerships that form the grid of tourism and tourist experience at a site.

CRISSCROSSING TOTAL SOCIAL FACTS: TRAVELOGICAL ANALYSIS

I begin this ethnographic travelogue with an event, and from this event I track out to a variety of crosscutting, intersecting, and overlapping stories, performances, symbols, conflicts, images, texts, and metaphors. This tracking is guided by the methodological implications of Mauss's (2016 [1925]) concept of total social fact, developed in the study of gift exchange and reciprocity. Mauss demonstrated that a given social fact (such as a gift, institution, site, or practice) is simultaneously a "thing" in multiple registers—for example, legal, political, personal, kinship, economic, and religious.[1] Thus to explain and understand any given fact requires a method of tracking these different facets and aspects that crisscross at least a few of these registers of meaning and contestation.[2]

All tourism destinations are total social facts: Chichén, like any place, is not just one, singular, unitary, homogenous place, but simultaneously a diversity and plurality of places in the same space.[3] Chichén Itzá is simultaneously an archaeological site, a tourist zone, a workplace, a subjective experience for thousands of locals and a million tourists a year, private property, collectively owned communal property, a symbol and representation of multiple quite different identities (Mexican, Yucatec, contemporary Maya, pre-Columbian Maya), and a sacred site of pilgrimage for pre-Columbian Maya and for twenty-first-century New Age spiritualists. In this ethnography, the initial point of entry to this crisscrossed entanglement of social facts is to understand Chichén as a platform and venue for politics, which in turn leads us to analyze its role as an economic resource, the laws by which it is owned, and ascriptions and experiences of being Maya.

The first destination of this ethnographic travelogue is the specific event of the selection of "Chichén" as one of the New Seven Wonders of the World, also known as N7W, on July 7, 2007. Similar to a dialogical approach that focuses on telling stories, we start with the text of a story in which the experiences of that day are described. We then start unraveling some of the different threads of politics, economics, identity, and history that are interwoven in these events and stories. We travel or track back, zigzagging to previous and subsequent historical moments, crisscrossing times, practices, performances, discourses, texts, and events.

CHICHÉN ITZÁ, JULY 7, 2007: A MARVELOUS PARTY AND PROTESTS

Imagine yourself there, broiling in the humid sun of Yucatán at 40° C (104° F), along with 8,650 tourists from México, the USA, and many other countries, who came in festival mode to experience the final worldwide voting for the N7W:

> Some celebrated the designation of Chichén Itzá as one of the modern Seven Wonders [*maravillas*] of the World; . . . others denounced the discrimination that converts living Maya today into objects. While the authorities and tourists partied, . . . the dozens of Maya artisans protesting a few meters away from the Pyramid of Kukulkán[4] [protested] that the Indigenous people have never been respected and they are kept in perpetual underdevelopment [*rezago*] and poverty.[5]

The thousands that had gathered in the main plaza were mostly Mexican and Yucatecan, since the international tourist high season had barely started. They came to hear the final count of the open vote for the New Seven Wonders of the World, which was being broadcast live in the plaza from Lisbon, Portugal.[6] TV Azteca, DIRECTV channel 420, Yucatán State Government Canal Trece TV, and "Canal del Sur" broadcast the long-anticipated event live throughout most of México and parts of the USA and South America. This was an event fit for a spring equinox at Chichén Itzá, which since 1984 has been a major state-sponsored tourism festival attraction (Castañeda 1996, especially 175–203; Himpele and Castañeda 1997). Live music by Orquesta Típica Yucalpetén and other regional artistic performances were interspersed with the usual lineup of speeches by politicians and archaeologists. The N7W voting in July 2007 coincided with the final month of political campaigning for the governor of the State of Yucatán; the candidate for the Institutional Revolutionary Party, or PRI, Ivonne Ortega Pacheco, was therefore present and was the most significant speaker, second only to the N7W live news feed from Lisbon.

México—that is, the government, citizens, and corporate businesses—recognized an opportunity as early as 2003, when "Chichén" was repeatedly voted onto shorter and shorter lists of competitors pending the final vote on July 7, 2007. To clarify, the N7W competition was not for entire sites, but for the most significant feature of these famous contenders. Thus there was often a conflation in popular understanding of the vote being for an entire city or place, such as "Chichén Itzá" versus a specific architectural feature, the iconic pyramid of K'uk'ulkan. A marketing alliance between the private and public sectors spawned a massive campaign to motivate Mexicans to vote for Chichén at least once, if not several times. For example, Coca-Cola distributed "Vote for Chichén" Coke cans, and Telmex issued phone recharge cards with an image of the Castillo (Chichén's main pyramid) and also subsidized votes made by cell phone calls and texts (see Castañeda 2009b, 294; figure 9.1). While global popular opinion supported the N7W movement, in México there was a definite but minor voice of disapproval from academics and governmental and international nongovernmental sectors about both N7W and Chichén as a winner.

N7W was created by the independent entrepreneur and adventurer Bernard Weber in 2000. Although UNESCO was initially involved with N7W, it soon disassociated itself from Weber and the nonprofit foundation. The issue was clear: N7W operated on a populist agenda of allowing anyone, anywhere, the determining role in selecting candidates via open, uncontrolled voting on the N7W website. In contrast, UNESCO, as mandated by its organizational structure and mission, regulates and supervises the list of World Heritage Sites in accordance with conventions established by and for member states; states and their politics therefore play the determinative role in these designations.

Weber established N7W with the aim of creating a popularly defined list of spectacular human heritage, modeled on the concept of the Seven Wonders of the Ancient World, that would circumvent the political process of states and international governmental agencies.[7] The Seven Wonders of the Ancient World is a contemporary composite of lists proposed in the Roman era of the most significant "wonders" created by peoples—Egyptians, Greeks, Persians, Babylonians—that Rome had colonized and that a civilized Roman therefore should see when traveling the empire (Perrotet 2002). The ancient lists identified significantly impressive architectural constructions, complexes, or features, but not entire cities. Thus, the modern list Weber sought to create followed

Figure 9.1. Coke can with "Vote for Chichén" label.
Photo credit: Quetzil E. Castañeda.

this precedent. In contrast to UNESCO regulations, N7W defined a seven-year process by which everyone in the world could vote (more than once, if they desired) in a populist, open-vote "election" that was purposefully outside of the objectives and regulatory determinations of national and international governmental organizations. UNESCO World Heritages Sites, on the other hand, are preponderantly entire sections of cities and settlements, or wide-ranging architectural complexes extending beyond a single building or monument. Significantly, N7W candidates were by definition only single buildings, monuments, or works of art—not whole sites or extensive sections of such.

In México a handful of academic experts, public intellectuals, and bureaucrats voiced critiques and lamentations. As reported in the news, these included protestations such as: "scientific criteria of selection were not used" (as if scientific criteria had ever been used, had existed, or could be created!); carrying capacity and other "impact analyses had not been conducted"; "it's not just one pyramid, but all of Chichén is a *maravilla*"; there are "thousands of other pre-Columbian Mexican sites that are '*maravillas*' because of 'ancient hydraulics.'" Critics contended that N7W had created a global profit-making and marketing scheme that benefited corporations while exploiting communities and heritage.[8] More interesting for us are the subaltern-styled groups[9] who protested N7W voting: "Chichén Itzá is not just the pyramid of Kukulkán: [it is] us, the peasants [*campesinos*], the artisans [*artesanos*], the field hands, workers, our culture, our language. . . . We are tired of the giant corporations that profit from us and the great cultural works left to us by our ancestors, asserted *artesano* leaders."[10]

First, to locate ourselves: the artisan leaders are all from the Maya community of Pisté (population roughly seven thousand). Historically speaking this settlement was part of the pre-Columbian metropolis of Chichén Itzá. Today Pisté is the primary center for all services in relation to Chichén tourism and, therefore, the socioeconomic core of the surrounding microregion of Maya communities. I have argued at length that Chichén and Pisté should not be considered two separate entities; rather, together they form the integrated tourism destination of Chichén, or what I have termed the Pisté-Chichén tourism complex (Castañeda 1996; 2001; 2009b; see also Breglia 2006; Armstrong-Fumero 2013).

Second, I consider the laments expressed by the news reports to be dog-whistle rhetoric because they index, embody, and perform dominant ideologies in Mexican nationalism that identify Indians (*Indígenas*), artisans (*artesanos*), workers (*trabajadores*), peasants (*campesinos*), and the *ejidatarios* (a special subset of peasants discussed below) as the great historical protagonists of the Mexican Revolution (1910–1917) and benefactors of the resulting 1917 Constitution that brought about the modern Mexican nation.

These protagonists became symbols of the legitimate right to rule, ideological identities of class belonging within the nation, and legal entitlement to and ownership of Mexican heritage. Mexican politics—the state, political parties, policies, and electoral campaigns—are organized around these ideological armatures. They form the lingua franca that mobilizes affects and other sentiments of national belonging and alliances of sociopolitical mobilization.

Resulting from the Mexican Revolution, these identity roles—peasant, Indian, worker, artisan, woman, *ejidatario*—became the ideologically proper citizens of the Mexican state. On the one hand, the Mexican Revolution forged the legitimacy of the state to the extent that it would govern in the name of, in service of, and for the benefit of these previously disfranchised citizens. In the course of the twentieth century these categories were converted to symbolic imaginaries and discourses that generated ideological roles and discourses of political action, through which the Mexican state governs civil and political society. On the other hand, individual citizens and citizen groups seeking to engage in political action to petition the state in social movements are compelled to claim these ideological identities, accurately or not, in order to assert their legitimate right to *interpellate*—that is, to name, call on, and demand action from—the state to uphold its role as the democratic government that serves, protects, and provides benefits to all citizens, especially these historical role types. When the Mexican state fails to work for the benefit of these citizens, sociopolitical crises erupt, as in the 1980s, when collapse of the peso led to a legitimacy crisis of the ruling political party and the contested presidential election in 1988. A national and state government strategy that led out of the crises of the 1980s was the increased development of tourism, in particular, in the Merida-Cancún tourism region.

PERFORMING IDEOLOGICAL CITIZENSHIP: THE PROPER ROLES AND IDENTITIES OF POLITICS

Mexican politics, especially in rural contexts, occurs through these citizenship identity roles, which therefore become sites of contestation. For example, the state has a wide array of welfare programs that target these citizenship groups, so persons try to assume these identities in order to be recipients. Despite the newspaper report quoted above, the politics of handicraft selling at Chichén involves a complicated negotiation of ideological citizenship. Tourism development of Cancún in the nearby state of Quintana Roo in the 1970s stimulated businesses and the State of Yucatán to begin to develop tourism in response to the skyrocketing number of visitors in the 1980s. In relation to this, peasants from Pisté began to work part of the year in the production and sale of handicrafts to tourists at Chichén because local hotel and restaurant service labor was closed off to them. Thus began the "invasions" of Chichén Itzá to sell inside the zone, in contradiction to laws that made this illegal use of heritage space (Castañeda 1996; 1998; 2009b; Breglia 2016; Armstrong-Fumero 2013).

Handicraft *vendedores* ("vendors") and *artesanos* ("artisans") are primarily from Pisté, and Pisté vendors have always occupied the leadership positions of this group, which has ranged in number from twenty-four in 1982 and seven hundred in 1987 to more than a thousand ten years into the third invasion (2005 to present). In the initial two invasions, 1982–1987 and 1994–1998, the Pisté leadership claimed to be *artesanos*, *campesinos*, and *ejidatarios* with the legitimate right to be inside the archaeological-tourist zone selling handicrafts. Federal law defines *ejidatarios* as the benefactors of the federal land grants (*ejidos*) that were given to peasant communities to be used in communal ownership and controlled by collective authority (*ejido*). *Ejidatarios* therefore have the right to use their *ejido* as they wish for their economic livelihood, but as discussed by Lisa Breglia (2006), this right of heritage conflicts with other archaeological heritage laws.

The majority of the initial twenty-four handicraft vendors were *ejidatarios* and thus *campesinos*, and most were *artesanos*. However, the exponential increase of vendors at Chichén quickly outnumbered the total number of *ejidatarios* in Pisté and nearby communities (approximately five hundred in 1984 and nine hundred by 1987). Further, the majority of these new vendors did not have occupational histories that included agricultural work. At all public events, in news articles, and in interactions with the state authorities, all the invaders prominently assert that they are *campesinos*, *artesanos*, and *ejidatarios* even though in objective economic terms this is unlikely to be accurate for the vast majority. Not only is the category of peasant problematic in the first place (Kearney 1996), but the majority of vendors (as opposed to artisans) long ago left agricultural labor for work in construction and an array of other service labor. By extension, news reporters who oppose the Pisté *vendedores* and *artesanos* always denounce them in three ways, by asserting that they are not *artesanos* and *campesinos*, that they are actually "wealthy" middle-class vendors, and that they are neither Maya nor even Indigenous.

The politics of Maya identity are also complicated. In brief, institutional racism in Yucatán since the mid-nineteenth century has resulted in a majority of Yucatán's population, to whom Maya identity is ascribed by Mexicans, Yucatecos, and foreigners, rejecting "Maya" as a label of self-identity. However, the development of mass international tourism based in Cancún and Maya Riviera since the 1990s has created new economic opportunities, social statuses, and these benefits that have increasingly motivated people to strategically self-identify as "Maya," especially in contexts involving foreigners; this is exquisitely discussed by Juan Castillo Cocom (2005; 2007) in his analysis of ethnoexodus (Castillo Cocom, Rodriguez, and Ashenbrener 2017; Castillo Cocom and Ríos Luviano 2012). In Pisté and communities surrounding Chichén, persons mostly reject identifying as Maya (but not to "being" Maya) because the identity is associated with economic impoverishment, educational ignorance, backwardness, and racial discrimination. These ascriptions and prejudices are generally absent with foreigners, or mostly present in the romanticist exoticization of the "Maya" as a great civilization (Castañeda 2004b; Castillo Cocom 2005; 2007; Castillo Cocom and Ríos Luviano 2012).

The vendors' strategy of identifying as "Maya" began in April of 2005 when their leadership initiated an alliance with the Zapatista Army of National Liberation (EZLN) and its spokesperson-leader, Subcomandante Marcos. This marked a shift from the previous vendor politics whereby the *vendedores* claimed to be *campesinos*, *artesanos*, and *ejidatarios*. As *ejidatarios* the handicraft vendors asserted the legal right to occupy and use their *ejido* lands, which include the ruins of Chichén. The ancient city extends at least five kilometers in all directions from the main plaza and thus overlays several nearby modern communities. The ruins restored as a tourism destination exist in the *ejido* lands of three different communities (Pisté, Xcalacoop, and San Francisco). While the lands are owned by the communities, the physical buildings and artifacts—the ruins themselves—are not.

Ruins—that is, archaeological heritage in the form of portable objects and immobile structures—are owned by the federal government as national patrimony in a form of stewardship. However, the lands in which the ruins exist are also juridically defined, and physically fenced off, by additional sets of laws that create conflicting and discontinuous juridical spaces: Chichén Itzá is simultaneously private property owned by the Barbachano family (an economic elite whose power dates to before the 1847 Caste War), an archaeological site of national patrimony administered in stewardship by the Mexican National Institute of Anthropology and History (INAH), and a tourism destination zone for visitors. In these juridical places it is illegal to sell handicrafts without special state or federal permits. Thus, the word "invasion" is itself a contested site. Used in news media, public understanding, and governmental discourse, it designates entering *la Zona Arqueológica* to sell handicrafts as an illegal act and effectively erases and silences it as a legal act of going to work on rightfully owned and used *ejido* lands. The term *la Zona*, however, refers to the proper and legally defined archaeological zone as well as to the ambiguously bounded, porous zone of tourism and the legally measured but sociologically ambivalent intersection of *ejido* lands that overlap in the space of these other two geographic entities. The vendors simply say, "*estamos en la zona*," or "*en terrenos ejidales*" ("we are in the zone," or "in *ejido* lands"), or "*entramos*" ("we entered") to refer to present or past actions as juridically rightful privileges.

It is now possible to understand the irony and ideological warfare of the newspaper article about N7W quoted above. The author, Rafael Mis Cobá, converts the negative rhetoric of the previous two years into ideological heroism. Instead of describing the vendors as middle-class, non-Indian merchants who had invaded the zone, Mis Cobá reports that local *artesanos*—implicitly coded as *campesinos* and Indigenous—are protesting "big corporations" that profit from Chichén. The Pisté *artesanos* are thus described as national heroes fighting the imperialism of globalization and capitalism for the rights and in the name of all Mexicans.

CACOPHONY AND PERFORMATIVITY OF "TOURIST ZERO"

Performativity refers to the capacity of words, speech, stories, written language, and dialogue—not just actions—to create reality. Performativity is the power of creating the world by performing, enacting, and expressing words, dialogues, stories, performances, interactions, and feelings.[11] Cacophony is a chaotic overload of diverse, disjunctive, dissonant, and deafening noises, mostly senseless, that creates confusion. Cacophony is the extravagant destruction of order, meaning, and coherence by sonic overload. Performativity is the creation of order by the enactment, practice, and performance of discourse. Yet at Chichén, the performance of cacophony is extraordinary: at some mysterious threshold it begins to convert senselessness and the erasure of meaning into another, extra dimension of order, coherence, meaning, and extraordinarily routine normal. In this way, contested meanings are rendered invisible and silent.

On January 1, 2006, Subcomandante Marcos initiated the "Other Campaign" in response to México's 2006 presidential election. This was a political tour of México organized in a fashion akin to an election campaign. Marcos's objective, however, was not to promote a candidate, but to disseminate the EZLN declarations, positions, and goals, as expressed in their 2005 Sixth Declaration of the Lacandon Forest.[12] Conspiracist critics argued that another, unstated objective was to debilitate populist candidate Andrés Manuel López Obrador, nicknamed AMLO (then running for president with the Party of the Democratic Revolution, or PRD), who was viewed as "not left enough" by the progressive purists among the new social movement advocates (Ross 2006, 301–78; Khasnabish 2013; Muñoz Ramírez 2008;

Hernández Navarro 2008; Vergara-Camus 2014).[13] Regardless of that debate, Marcos assumed the self-identification of *Delegado Zero* (Candidate Zero)—that is, the "Not a Candidate"—since he was campaigning not to be elected but to popularize the EZLN sociopolitical agenda by giving speeches at every "not-a-campaign" stop throughout México.

Following the end of the Cold War and of communism as an antidote or opposition to capitalism, the EZLN used wooden guns to assault and take over the city of San Cristóbal de las Casas, Chiapas, México, on January 1, 1994, in a symbolic performance of revolt (if not revolution) against neoliberal capitalism and globalization. In this fashion, the EZLN and their "non-leader leader"—that is, Subcomandante Marcos—created an historic landmark date for the beginning of "new social movements," or democratic alliances of diverse social groups seeking what is generally referred to as social and economic justice. The EZLN was a globally popular movement that generated a worldwide following, as well as romanticized adoration from those pursuing the cause against globalization.

Ironically, however, the EZLN created a new form of alternative neoliberal tourism, called Zapatista tourism or Zapa-tourism. This "activist tourism" is anti-neoliberal in ideology and thoroughly neoliberal in practice (Babb 2010; Heyns 2014; Nelson 2009; Pitarch 2004). Zapa-tourism not only includes visits to "radical-chic tourist attractions" (Nelson 2009) showcasing everyday life in Zapatista towns, but also converts EZLN media events, community ritual, political speeches and declarations, documentary filming, and news interviews with Marcos into tourism spectacles. The EZLN knowingly converted itself into a tourism performance as part of its strategy to generate global support. Marcos himself became the main tourism attraction and destination for this activist tourism. Therefore, it is appropriate to assume the ironic, if not satirical, voice typical of Latin American politics and the humor of popular social criticism: when Subcomandante Marcos hit the Other Campaign trail as Delegate Zero, he also performed his ideological role as the voice and mask against neoliberalism that we can call "Tourist Zero." Given the interchangeability and overlap of tourism roles (E. Bruner 1995; Himpele and Castañeda 1997), we should not be surprised that Tourist Zero, like a celebrity who performs a stage persona, is a tourist and a mobile tourist destination.

JULY 27, 2006: "WE WILL DEFEND THEIR CULTURE"

When he is traveling celebrity-style, the international press fawns and fights over Subcomandante Marcos. He arrived in Merida to plot strategies with allies, with an entourage of followers as expected, but he was received with acerbic, satirical critique in the press. While Marcos fed the reporters his news, the press turned its criticism-cum-reporting to the global news about Chichén as one of the final candidates for New Seven Wonders of the World. The general anti-Marcos sentiments I encountered in Pisté in the early 2000s (Castañeda 2005a) were forgotten and muted during the time the Pisté vendor union representatives and lawyers developed an alliance with the EZLN, starting in spring 2005. Chichén was an ideal venue for his then-anticipated political campaigning as Delegate Zero, and these alliance-building exchanges created a positive grassroots reception with the vendors in the archaeological zone. A two-way exchange tour was arranged, with the Pisté leadership visiting EZLN headquarters in late 2005 and spring 2006, and then hosting Subcomandante Marcos's visit to Chichén on July 27, 2006, the anniversary of the Caste War of Yucatán. On this day, Marcos's discourse included the following:

> In all parts where there are signs of our ancestors, Palenque, Bonampak, Chichén Itzá, El Tajín, Monte Alban, Teotihuacan, there are Indigenous brothers that are rising up to defend the blood that we inherited from the ancestors. . . . The bad government wants to make our history something dead, as if it were an animal, and sell it, but we have to unite to prevent this! . . . We will defend their culture, their language, their ways of dressing, their music, their dance, and their forms of love until the end of the road.[14]

The rhetoric of an inclusive *we* of Mexican mestizos called to defend "them"—the *Indígenas*—is an ideological and racial othering. The racialized *Indígenas*, "them," are identified as ideological citizens in need of help. In an extravagant comment, a seeming non-sequitur, "they" even need help defending their "forms of love." This extravagant wandering way off topic is a gratuitous show of hubris; without apparent sense or meaning, the sonic detritus confuses and distracts. While this confusion of contested meaning is heard clamoring, other flagrant incongruities in Marcos's touristic performance of politics reveals ideologies at war: Marcos's invocation of ideological Indigeneity imagined in the romanticist rhetorics of anticapitalism and anti-neoliberalism is at odds, first, with the normative cultural politics of self-identification as Indigenous and as Maya in Yucatán and, second, with the explicitly commercial-capitalist objectives of local Pisté Maya vendors and artisans. The defense of their forms of love contributes to silence and conflates local understandings in a polyphonic chorus (E. Bruner 1993b, 14–15; Castañeda 2004b; 2005a; 2009a).[15] Tourist Zero's discourse explicitly points to conflicting visions of society, but the words and performance silence and disappear other struggles, including the vision and goals of the Pisté vendor associates.

Delegate Zero performs himself—Subcomandante Marcos—at Chichén, which constitutes a theater, mirror, and echo chamber for his spectacle of self-(re)presentation. Delegate Zero is Tourist Zero, who experiences Chichén as a mirror reflecting the truisms of his own discourses and previously established understandings. Either we assume Marcos as a one-dimensional performance of his ideological celebrity persona, or we suspect that Marcos is only cynically enacting a discourse of ignorance that expresses no relationship of intimacy or empathy with local perspectives, no grasp of the specificity of the local social history and politics of Chichén, and no understanding of Pisté vendors as petty capitalists committed to globalization. His rhetoric only acknowledges "Maya" *campesinos* and *artesanos* as prescribed by ideologies of both the Mexican state and his anti-neoliberal leftist Indigenism.

Marcos continued, according to the news reports, "The government wants to convert [Chichén Itzá] into a commercial center and wants to eliminate that which dirties it and makes it ugly, and for that reason it wants to eliminate the indigenous Maya *artesanos* so that only whites can walk on the dead."[16] Yes, between 1923 and 1941, Mexican and US archaeologists, using Pisté labor, built this space to be a commercial center for tourism, or what Alfred V. Kidder called a "Mecca" of tourism (Castañeda 1996; 2005b; Castañeda and Mathews 2013). According to the head archaeologist, the goals were to "make it an enduring monument to the genius of the ancient Maya, . . . [and] if Chichén can be kept both interesting and beautiful, it will without question become a Mecca of travel, and incidentally, a most valuable asset for archaeology" (Kidder 1930, 98–99; Castañeda 1996; 2010; Castañeda and Mathews 2013). Yes, the government does want to scrub the site clean of Indigenous artisans and vendors to keep it "interesting and beautiful." But, no, contrary to Marcos's declarations, the Maya *artesanos* and *vendedores* do want to be there precisely in order to commercialize and commoditize Chichén Itzá, Mexican heritage, and Maya civilization with the aim of maximizing their own profits and prospects for extracting economic value from ruins. They claim their right to *kaxtik u kuxtal*—"to seek one's livelihood"—as one sees fit (Castañeda 1996; 1998).

In this cacophonous performativity, do you also hear how the romanticized politicization of Maya identity as Indigenous by Marcos and the romanticist exoticization of Maya culture as a Lost Civilization by the tourism industry converge and meld smoothly, one into the other, into the same narrative of Maya otherness? Do you see the mass tourists and Tourist Zero sitting side-by-side in the plaza built by archaeologists in the 1920s, as they feast like cannibals on "the Maya"? But are "the Maya" unwitting victims, chewed up and strewn about by tourism-cannibalism? The press quoted Silvia Mex, the most powerful and enduring of several female leaders of vendors, who was able to unify all vendor factions in one union when she became its president soon after the invasion at the start of 2005. She retains this leadership position in 2018 even though the union has split into three unions. Listen to the loud and passionate words she used to motivate her fellow vendors and artisans: "The Mayas of today no longer have fear because we are dignified to be called 'Yuca-tercos' in this struggle. We are against bad government and there is no money or power that will hold us back. If we are going to die, it will not be for nothing. We are the new Maya Zapatistas."[17]

"Yucatercos are the new Maya Zapatistas," Silvia Mex says. "Yes, I am *terca* [meaning "stubborn," "pig-headed," "willfully ignorant," "uneducated"] and I am *Yuca* [a slur against the "provincial" Yucatecans by urbanite central Mexicans]." One might suspect that this is a reply to an insult thrown at her in the press or in private by a politician who sought to block the political objectives of the Pisté vendors, which is to have their right to sell inside the zone legitimated either by law or by a legally contracted agreement. Her reference to this insult and her defiant acceptance exemplifies the dialogical conflicts and contestation of meaning that occur in and through shadow dialogues and hidden, backstory interactions. In order to legitimize their ability to commercialize Chichén Itzá and Maya culture, the vendors needed to protest the corporate, capitalist, and state commercialization of the site—and to achieve this they needed to become "proper" Maya fitting the ideological prescriptions of the progressive left and the procapitalist right.

"BECOMING MAYA" (JULY 30, 2005)

Right after Christmas 2004, two and half years before the N7W vote, artisans and handicraft vendors from Pisté and nearby communities again entered Chichén Itzá. As on previous occasions, the vendors had mobilized with the goal of positioning themselves in the path of tourists inside the tourist zone to sustain their economic livelihood. In the two previous invasions (1982–1987 and 1994–1998), artisans and vendors had claimed their right to use their *ejido* in order to sell handicrafts to tourists (Breglia 2006; Castañeda 1996; 1998; 2009b). Within the first month of the third invasion they shifted their discourse of legitimacy to that of "being Maya" and thus having cultural property rights as Maya from nearby towns. Their strategy of being Maya was forged by the new artisan and vendor union leadership, who created an alliance with the EZLN and organized visits to Chiapas to meet with Marcos. The *artesanos* and *campesinos* were learning to become Maya, something they had previously resisted in various political mobilizations of Pisté (Castañeda 1996; 1998; 2009b).

In August 2005 I arrived in Chichén for a visit. A longtime friend greeted me and, unable to contain her contagious laughter, launched into an account of the great spectacle I had missed during my time away. All the *artesanos* and *vendedores* had participated in a protest event for their rights inside the zone on the 158th anniversary of the start of the Caste War (often characterized as a race war between whites and Indians).[18] "You should have been there, my friend," she repeated again and again, and she added with frisson: "*Pretendímos ser Maya*" ("We pretended, or intended, to be Maya"). In translation: We were consciously motivated to try to be and to perform being Maya. We attempted to be Maya and asserted Mayaness as our identity. Hers was not a little giggle but a full-throttle, scintillating laugh. In contrast to their prior political strategies for invading the zone, this was the first time they enacted and performed Maya identity politics. The self-conscious giddiness provoked by this performance of being Maya suggest a fundamental slippage: the Maya are "almost but not quite" ever Maya.[19]

Theatre and Spectacle: The Intensification of Space

Following a visit by the Pisté leadership to the EZLN in the summer of 2006, Subcomandante Marcos returned the visit to Chichén Itzá on January 21, 2006. The *Por Esto!* news article on Delegate Zero's visit includes a photo of Teresa, one of the original 1980s vendors and a leader, not laughing but crying, and walking out of Chichén Itzá with her arms around Marcos—Tourist Zero. Here, the Pisté merchants of Maya culture are not laughing but rather crying as they intentionally perform and enact being Maya ("*Pretendimos ser Maya*") to protest, without irony, the commercialization of Maya culture and of Chichén. This is pure spectacle, pure theatre (Castañeda 2001). Chichén is spectacular in all senses of the word—that is, referring to vision, visuality, the viewer, and thus also speculation or theory: the spectator is a theorist (which is etymologically derived from *viewer*) who speculates or theorizes the meanings and significance of sights, interactions, and performances. As travelogical ethnographers, we make sense of the total social fact by tracking and tracing elements of the event to other events, discourses, texts, contestations, and performances.

At Chichén, events of all types—artistic performances, social protest, New Age spiritualist ritual, sightseeing by Mexican or foreign politicians, state-sponsored tourist festivals, and campaign rallies of political parties—continually occur. These events, like the 2005 protest-anniversary event, the 2006 Marcos visit, and the 2007 N7W vote, typically consist of the same spectacle of performances seen at the site's annual spring equinox festival: music, poetry, singing, traditional Yucatec Jarana dance, TV camera crews, news media, speeches, and banners, all staged on a platform in the main plaza with seven to ten thousand sightseers. "Chichén," especially the main plaza and the Pyramid of K'uk'ulkan, is produced as and itself staged as a stage, a palimpsest, a mise-en-scène, and an open-air theatre for the events and the performance of assorted stories (discourses, ideologies, values) in guises of political conflicts, culture, tourist consumption, anthropology, and spiritualism. Chichén Itzá is constructed and reconstructed by all variety of actors and agents (archaeologists, politicians, tourism businesses, handicraft makers, and producers) as a space of and for contestation, which is itself a dialogical oscillation between cacophony and performativity.

MARCH 29, 2010: USUFRUCT, PRIVATE PROPERTY, AND SENTIMENTAL OWNERSHIP

> A historic transaction between the Government of Yucatán and businessman Hans Jurgen Thies Barbachano yesterday [March 29, 2010] allowed the State to buy 83 hectares of land where the archaeological site of Chichén Itza sits for 220 million pesos (US $17,800,150).[20]

There is never anything natural about how a site is defined in extension, boundaries, and attributes. Every place is created or constructed—that is, defined and experienced by social criteria that constitute the meaning and significance of space and places. The term *la zona*, or "the zone," is not only literally defined in different laws by intragovernmental agreements and conventions between private owners and government agencies, but also metaphorically defined by competing sets of amorphous, porous, symbolic, and sentimental meanings and values that comprise the ruins as national patrimony, pre-Columbian civilization, workplace, museum, and Maya culture. This complexity can be descriptively reduced to three primary types of contested "heritage" space: federal patrimony, private property, and community *ejido* land (Breglia 2006; 2016).[21]

The state, as federal steward, "owns" the ruins, or monuments (that is, the tangible architectural buildings, arrangements, and objects) created by pre-Columbian Indigenous peoples such as the Maya. However, the Mexican government does not own the land in which this national patrimony exists. Instead, the land in which the ruins exist is owned either as private property (by the Barbachano family, a Yucatán oligarchic family) or as communal *ejido* lands (by the communities of Pisté and Xcalacoop; Castañeda 1996; 1998; 2009b; Breglia 2006; 2016; Albright 2015). In 1985 the state government of Yucatán negotiated a legal arrangement with the federal government that granted the State of Yucatán the right to operate and administer all "touristic services" at all of the state's archaeological-tourist sites. The physical site itself is therefore a contested space carved up and controlled by different, overlapping, and competing registers of sentimental, traditional, and juridical ownerships, private and collective property rights, contracted and communal usufruct, and rental conventions.

The July 30, 2006 Pisté vendors' protest speech, called the *Declaración y Protesta en el 158th Aniversario de la Guerra de Castas*, enacted a shift in how *artesanos* legitimized their presence inside the tourist zone of Chichén. They dropped their claims as *ejidatarios* to instead claim ownership of the site as Maya from the nearby town of Pisté, in accordance with customary law. But neither this startling shift to a strategic essentialism nor Delegate Zero's January 2006 tour of Chichén led to a broad public understanding that the actual site of Chichén, although national patrimony, was not owned by the Mexican state. Rather it took the N7W vote for this knowledge to begin to circulate among ever-widening Mexican publics. Although the issue was generally ignored in the five-year marketing campaign leading up to July 7, 2007, recognition as a N7W opened the floodgate of critique. It was scandalous for Mexicans to learn that Chichén was private, not public, property. By 2010, there was nonstop outcry in national news and public media against this de facto, if not legal, privatization of Chichén as national patrimony.

Thus, our story returns to the beginning: on Chichén's plaza on July 7, 2007, Ivonne Ortega Pacheco appeared as the PRI Yucatán gubernatorial candidate, promising to resolve Chichén Itzá's twenty-five-year-old problems. She referenced not landownership issues but the "invasion" problem. No one really knew if her plans extended beyond powerful campaign rhetoric. Then, out of the blue, the state government of Yucatán purchased parts of Chichén Itzá from the Barbachano family in March 2010. In the three years between the campaign promise and this state purchase, the various pressures, meetings, movements, and attempts by the federal—not state—government to buy the Barbachano land holdings that had failed. On the basis of publicly available news, it is difficult to know the nature of the obstacles; it seems that the amount of payment for the expropriation, the usufruct rights to be retained, and the actual spaces to be expropriated were the most difficult issues.

At one point a member of the Barbachano family asserted that the deceased patriarch, Don Fernando, had already ceded the rights or transferred legal ownership of the crucial land on which the monumental heritage exists to the INAH. Needless to say this document has never been located, if indeed anyone has searched for it. Thus it is an open and public question that has increasingly been disseminated and debated in México: whether the federal government or Barbachano family members owned which specific surface lands—and where and with what boundaries and dimensions. This juridical and epistemological uncertainty is part of what Breglia refers to as the "monumental ambivalence" of Chichén, which allowed Mexican laws of usufruct and possession to take precedence.[22]

In March 2010, three years after both the N7W voting and her election as Yucatec governor, Ortega Pacheco announced the state government's purchase of a major portion of the Barbachano-family-owned property containing Chichén ruins. This expropriation left specific land, such as hotel properties, in their ownership but did not include the purchase of Pisté *ejido* lands. Not only did Pisté *ejidatarios* not sell their *ejido* land "in" Chichén (though other nearby communities had previously sold all or parts of their *ejido* grants), most of the nearly 1,200 vendors are not *ejidatarios* and have no direct usufruct rights. The ideological rhetoric of being peasants and workers was doomed to fail. Their gambit, initiated in 2005, of pretending, intending, performing, and "becoming" Maya paid off—in fact, it was the only way to achieve long-term positive results. An ongoing negotiation since the 1980s between the State of Yucatán and the various vendor union groups has resulted in the establishment of rules of behavior and protocols for selling inside the zone. This constitutes a self-disciplinization of *vendedores* and *artesanos* that fits Foucault's analysis of subjectivization, or how persons become proper citizens of the state through an integration of individual subjectivity to the given norms of the society. Today, fourteen years later, they are still inside the zone.

The story does not end here. The regional press, which is allied with big business of the region, is periodically publishing criticism against the vendors, except as noted. There is public discussion and fear that N7W status will be lost, ostensibly due to lack of compliance with the N7W conservation regulations, maintenance, order, and attention to preservation.[23] But all these reasons are linked in public opinion to a common underlying cause—the vendor invasion. This reiterates a moral discourse about "touristic image," that is, the *imagen turístico* experienced by tourists that has been used since the 1980s: those who opposed the vendor invasion have used and continue to use the metaphor of *mal imagen*, the "bad image" that Pisté vendors create for tourists, to politically delegitimate the vendors inside the zone (Castañeda 1996; 1998). In turn, the Pisté vendors have increasingly subordinated themselves in the name of producing a "good image" for tourists during their visits to Chichén. They assumed technologies of self to become proper subjects and citizen-vendors within la zona.

CONCLUSION

This chapter used an ethnographic storytelling style to analyze Chichén Itzá as a contested site of tourism. The ethnography is analytically focused on making sense of an event in terms of the overt and underlying meanings as conflicting and conflicted messages; this is neither an explanation of an event as an organized, structured activity, nor an interpretation of its meaning or the experience of it for specific actors. The event, N7W voting, itself is a story—one of several possible news representations of it—that assigns meaning to the event. Ethnography makes sense of a story of the event by tracking other words, texts, discourses, and performances about yet other stories and narratives. Tracking or traveling the analysis is concerned with meaning not for its own sake, but to reveal hidden or overt political messages, implications, effects, contradictions, and conflicts. The object of study is therefore neither the experience of tourists (or of ethnographers, guides, or locals), nor the meaning that they put into or get out of the event, storytelling, or performance. Instead, the object that is analyzed here is the ways different ethnographic (that is, social) facts crisscross, intersect, or careen away: political ideologies of social agency in México; the performance of Mexican electoral politics; heritage tourism; legalities of ownership; commodification of archaeology for tourist consumption; new social movements as manifested by the EZLN and performed by its leader Subcomandante Marcos; the politics of handicraft work; and the contradictions between external discourses of ascribed identities and embodied practices of being Maya.

The analysis moved backward and forward in time to crisscross these different kinds of ethnographic objects. The concept of total social fact does not imply or have the goal of a description of a completely unified, discretely bounded, total thing to be studied, whether this is conceived as an intangible abstraction—such as a culture, a community, or tourism—or as a materially tangible object of study—such as a tourism destination, an archaeological site, or a contested property. The total social fact concept refers—as do Bruner's concepts of contested sites, borderzones, and dialogical processes—to analytically unlimited extension of connection to other narratives, discourses, and practices. Thus these concepts identify connections between seemingly unrelated words, stories, and practices, and explain how they intersect in a given tourism site or destination.

What, then, is the significant difference between Bruner's approach and my own here? Differences between dialogical and textual approaches can be quite large in terms of the histories of these theories. Significant differences are better found in nuanced differences in the conceptualization of the ethnographic objects of study, the analytical method, and the objective or goal of ethnographic analysis. The dialogism of this chapter dovetails with Bruner, but points away from tourism per se, and toward encompassing politics and other ethnographic communities, as do other chapters in this volume. Bruner's own work was written at a time when the anthropology of tourism tended to focus on tourism interactions between social actors explicitly enacting tourism roles (tourist, guide, local, etc.). Bruner's work also focused on these situated interactions, but he most often developed concepts—such as tourism encounter and the borderzone—to implicitly critique earlier kinds of priorities and approaches. He accomplished this by studying these tourism interactions, places, and cultures in ways that directed anthropologists of tourism out of understanding tourism in a restricted sense of the activities and actors who engage in tourism. He decentered tourists (and other role types such as guide, host, and ethnographer) from the core, essential conceptualization of tourism, even as these were his ethnographic objects of analysis. He thus resituated tourism as a dialogical process of meaning-making that includes transnational dynamics and social institutions as much as the humanist experience and performance of cross-cultural exchange. Bruner's most significant contribution is the way in which he thereby led subsequent generations of anthropologists to do more robust ethnographies of tourism that increasingly located tourism as one among many other dynamics, processes, institutions, struggles, and ways of creating and contesting meanings.

NOTES

1. Mauss's concept builds on Émile Durkheim's idea of social facts. Although both spawned an array of diverse schools of thought, my use of Mauss's total social facts derives from Boon's teachings and crisscrossing anthropology (e.g., Boon 1982; 1999).

2. George Marcus (1995) proposes following or tracking different kinds of "ethnographic objects" or social facts as these manifest in or connect with diverse settings ("multiple sites"). This idea resonates with the interpretation of Mauss's concept of total social facts used here and has affinities with Bruner's notion of the borderzone, especially in relation to tracking transnational and international contexts and sites (e.g., E. Bruner 2005b, 13–17).

3. In my previous work (e.g., Castañeda 1996; 2001) on the contested meanings of Chichén Itzá, I developed the metaphors of palimpsest and stratigraphy as tools of analysis.

4. This is an example of a word being a contested site of different orthographies or rules of spelling and writing that track political hegemonies and power dynamics. The news article ironically reports on protests of colonial legacy while unaware of the power dynamics in the use of Mexican Spanish orthography (Kukulkán) versus Maya spellings, whether contemporary (K'uk'ulkan) or twentieth-century modernist (Kukulcan).

5. Rafael Mis Cobá, "Festejo y protesta," *Por Esto!*, July 8, 2007, www.poresto.net/content/view/7144/70/ (accessed July 9, 2007). All original Spanish language quotes, including newspaper articles, are translated by the author.

6. See "Es Chichén Itzá una de las nuevas maravillas," *El Universal*, July 7, 2007, http://archivo.eluniversal.com.mx/notas/435648.html (accessed December 28, 2017);

Rafael Gómez Chi, "Ignorantes e ineptos," *Por Esto!*, July 6, 2007, www.poresto.net/content/view/6796/70/ (accessed July 9, 2007); Rafael Gómez Chi, "Larga historia de la antigua finca de Chichén y sus anexas," *Por Esto!*, 2007, http://www.poresto.net/ver_nota.php?zona=yucatan&idSeccion=1&idTitulo=11773 (accessed January 2, 2018); Rafael Mis Cobá, "Festejo y protesta," *Por Esto!*, July 8, 2007.

7. There is debate as to why, when, and by whom the list of Seven Wonders of the Ancient World was created, as well as how and by whom it was used. Nonetheless, the seven wonders are fairly consistently identified as the Colossus of Rhodes, Hanging Gardens of Babylon, Temple of Artemis in Ephesus, Lighthouse of Alexandria, Pyramid of Giza, Mausoleum of Halicarnassus, and the Temple of Zeus at Olympia. All but the Pyramid of Giza have been destroyed.

8. See, e.g., "Es Chichén Itzá una de las nuevas maravillas," *El Universal*, July 7, 2007; Rafael Gómez Chi, "Ignorantes e ineptos," *Por Esto!*, July 6, 2007; Rafael Gómez Chi, "Larga historia de la antigua finca de Chichén y sus anexas," *Por Esto!*, 2007; Rafael Mis Cobá, "Festejo y protesta," *Por Esto!*, July 8, 2007.

9. *Subaltern* refers to an array of identity groups (e.g., ethnic-racial, sexual, linguistic, religious, cultural, or migratory minority) that have been subjugated by colonialism or are marginalized by state domination.

10. Rafael Mis Cobá, "Festejo y protesta," *Por Esto!*, July 8, 2007.

11. This concept is associated with scholars such as J. L. Austin, Judith Butler and Mikhail Bakhtin, among others working in a range of fields, such as ordinary language philosophy. It is a fundamental, core concept of most queer theory research.

12. "Sixth Declaration of the Selva Lacandona," Enlace-Zapatista, Zapatista Army of National Liberation (EZLN), 2005, http://enlacezapatista.ezln.org.mx/sdsl-en/ (accessed March 10, 2018).

13. The two dominant political parties in México, along with other, minor parties, signed an alliance pact before Christmas 2017 as a failed strategy to defeat AMLO in the July 1, 2018 Mexican presidential elections. Subcomandante Marcos, on the other hand, "took off the iconic mask" to retire the hero of anti-neoliberalism. See Alice Speri, "'Subcomandante Marcos No Longer Exists': Zapatista Leader Retires His Nom de Guerre," *Vice News*, May 26, 2014, https://news.vice.com/article/subcomandante-marcos-no-longer-exists-zapatista-leader-retires-his-nom-de-guerre (accessed December 30, 2018).

14. "Nadie nos va a sacar de Chichén," *Por Esto!*, January 21, 2006.

15. The words *indigeneity* and *indigenous* in their generic senses are not capitalized. In México, these words are capitalized because Indigenous peoples are sociopolitically and juridically defined communities with legal status. *Indigenism* refers to explicit state policy and ideologies that are implemented by specific governmental agencies.

16. "Nadie nos va a sacar de Chichén," *Por Esto!*, January 21, 2006.

17. Ibid.

18. The date of the Caste War is another site of contestation. Dominant Yucatec historiography holds that the war started on July 30, 1817, in the town of Tepich, when Maya leaders attacked the "whites" of this town. The nearby community of Tihosuco is currently reinventing itself and its heritage as the place of the war's

origin with a festival celebrating not the start of the war on July 30, but the "first spark" on the 26th, when "white" Yucatecos assassinated a Maya leader in Tihosuco. The goal, however, is not an Indigenist or identity politics project of rewriting history; instead, there are two intertwined objectives, Maya cultural revitalization fueled by the creation of Tihosuco as a tourist attraction feeding off the voluminous Cancún-Maya Riviera destination. See Campbell (2018); Alejandro Campo, "The 170th Anniversary of the Caste War," *The Playa Times*, August 19, 2017, www.theplayatimes.com/2017/08/19/170th-anniversary-caste-war/ (accessed December 30, 2017); "Tihosuco celebra a oscuras el 170° aniversario de la Guerra de Castas," *El Chilam Balam*, n.d., http://elchilambalam.com/tihosuco-guerra-de-castas/ (accessed December 30, 2017).

19. The quoted phrase was coined by Homi Bhabha (1994) to critique how British colonialism created implicit racial limitations to the sociopolitical opportunities of colonized peoples while using an explicit colonial policy of "lifting" the colonized out of "premodern" savagery.

20. "Mexican State of Yucatán Buys Archaeological Site of Chichen Itza from Private Landowner," *Art Daily*, March 30, 2010, http://artdaily.com/news/37171/Mexican-State-of-Yucatan-Buys-Archaeological-Site-of-Chichen-Itza-from-Private-Landowner (accessed December 30, 2017). See also "Chichén Itzá perdería el título de 'Maravilla del Mundo,'" *Diario de Yucatán*, November 15, 2016, http://yucatan.com.mx/merida/turismo/chichen-itza-perderia-titulo-maravilla-del-mundo (accessed December 28, 2017).

21. In common popular use, *patrimony* and *heritage* are essentially synonymous. As analytical concepts I employ *patrimony* to refer to things used by the state to create sentiments of national belonging (nationalism) and *heritage* to refer to practices of cultural groups that are not nation-states to assert both legitimacy as minorities within a nation and political control over their own resources (Castañeda 2009b; 2009c).

22. See Breglia (2006; 2016), Castañeda (1996; 2005a; 2009a; 2009b), and Armstrong-Fumero (2013) for ethnographic histories of the struggles to own, use, and control Chichén and other Maya heritage sites in Yucatán.

23. See "Buscan reconfirmar a Chichén Itzá como Maravilla del Mundo," *Proceso*, April 12, 2016, www.proceso.com.mx/436799/buscan-reconfirmar-a-chichen-itza-maravilla-del-mundo (accessed December 28, 2017); Mathieu Tourliere, "Chichén Itzá, en riesgo de perder la categoría de 'maravilla del mundo,'" *Proceso*, November 13, 2016, www.proceso.com.mx/462329/chichen-itza-en-riesgo-perder-la-categoria-maravilla-mundo (accessed December 28, 2017); "Castillo de Chichén dejaría de ser Maravilla del Mundo," *Mileno Novedades*, Yucatán section, November 16, 2016, https://sipse.com/milenio/chichen-itza-kukulcan-riesgo-categoria-patrimonio-humanidad-unesco-maravilla-230344.html (accessed January 2, 2018).

Part V

HOMECOMINGS AND FUTURE DIRECTIONS

Chapter Ten

Taking Tourism Seriously

A Conversation with Edward M. Bruner

Michael A. Di Giovine

Editors' Note: This chapter takes the form of a single conversation between Edward Bruner and Michael Di Giovine. However, in actuality it is based on two periods of interviewing over the span of two years, totaling in all around six hours. The first was held as a free-flowing, recorded conversation between Edward Bruner and Michael Di Giovine, Naomi Leite, and Quetzil Castaneda, held in Bruner's hotel room during the American Anthropological Association's 2014 meeting in Washington, DC.[1] A year later Di Giovine met up with Bruner in Chicago for a more formal set of interviews held over a two-day period, drawing on a list of topics and questions he and Leite had developed beforehand. He then edited the resulting transcripts into the present form, in collaboration with Bruner, who reviewed the document and clarified certain points.[2] Thus, because this was a collaborative effort among several people, questions are not attributed to specific individuals. And while both the questions and answers have been synthesized and lightly edited, original wording has been maintained wherever possible.

One title the editors considered for this volume was *Edward Bruner and Interpretive Theory in the Anthropology of Tourism.* As a self-described "restless anthropologist" (E. Bruner 2012) who has moved field sites and topics of inquiry several times in your long career, do you see interpretive theory in relation to tourism as your primary contribution to anthropology?

Most anthropologists would probably regard that as my major contribution and it may be correct. I had a retirement party at Illinois in 1994, and Barbara Babcock, Jim Fernandez, and the sociologist Norm Denzin came to speak. They talked about my work beginning in the 1980s. After all, I got my PhD in 1954, and then studied American Indians and then did research in Sumatra.

This earlier work was not mentioned by the speakers. They omitted discussion of the first twenty-six years of my career. The pattern was repeated at a symposium in 2005 at the American Anthropological Association devoted to my work and contributions. Dean MacCannell and Barbara Kirshenblatt-Gimblett were among those who gave papers at the symposium.

How did that make you feel?
I came to understand it. My early studies were done in a positivistic, non-humanistic, politically insensitive era. In the late 1960s and certainly by the 1980s we had a major paradigm shift in anthropology. It is difficult to fit old studies into a new paradigm. It raised for me an interesting question about anthropological memory.

However, there was one exception to this at the 2005 American Anthropological Association, in the symposium on my work. Nelson Graburn was in the audience but not one of the speakers on the program. After the speakers finished he raised his hand, moved up to the podium, and delivered a paper he had written on the plane en route to the symposium. It was unexpected. He talked about my earlier work, particularly where I took a constructivist view of kinship rather than seeing kinship as following fixed roles, as if you just fit in the slot. Kinship systems were then seen as more formal and mechanical, and we did not get many analyses of how people constructed their kinship roles. For example, one Indian woman complained about a girl who called her "auntie," the white way, but in the Indian way she would be "mother." The woman wanted to be called mother and said the children in that family all acted like white kids. People had choices, and they constructed kinship roles depending on how they saw their situation. Graburn felt that my work was a fulcrum and a turning point in moving from a positivistic to a constructivist approach in kinship studies, but nobody recognized it at the time. Truthfully, I myself didn't recognize it at the time, or possibly I was not bold enough to criticize prevailing approaches and contrast them with my own.[3]

Would you see constructivism as a major thread running through your work, then?
Oh, yes.

So how did you get to this point? What books or influences led to your current constructivist position?
I'll go through it as briefly as I can, but you feel free to ask me to go into more depth on any aspect.

At Ohio State graduate school, from 1948 to 1950, I became close to Donald Campbell, a distinguished social psychologist, who introduced me to a book by Frederic Bartlett called *Remembering: A Study in Experimental and Social Psychology*, published in 1932 in Cambridge. It was a remarkably influential book and is still in print. The key idea was that remembering was more of a construction than a reproduction. Memories are not just stored up, fixed, unchanged, timeless, and then recalled at a later date. A memory is not just recalling an exact copy of the original. Memories and retellings are always of the moment, based on previous experiences, the audience, and current circumstances. Recent neuroscience has confirmed this position.

In Chicago, 1950–1954, my first class in social anthropology was taught by Robert Redfield. It was the basic graduate course for all entering students. One book we read was *Mind, Self, and Society* by George Herbert Mead (1934). It was the founding text in symbolic interactionism, and Redfield had included it in the social anthropology graduate class. I learned there that persons have active selves that construct roles rather than merely occupy them. Roles or kinship positions, so prominent in the anthropology of the time, are not fixed slots, mechanically performed, but are constructed. In Pavlovian theory, stimulus leads to a response (S-R); in symbolic interactionism, stimulus leads to an act of interpretation which then leads to a response (S-I-R).

Other influences followed. I read the European poststructuralists and other European scholars in a faculty seminar at Illinois beginning in 1980 that lasted for decades. I was also very influenced by the Russian linguist Mikhail Bakhtin, who advocated that language is alive, uttered in dialogic interplay rather than as a static monologist entity.[4] He was opposed to structural linguistics, then the dominant theory. All stories for Bakhtin are told in voices and are polyphonic, always in interaction at any historical moment.

I was the faculty director in a study abroad program in 1983–1984. As a co-faculty member I selected the folklorist Barbara Kirshenblatt-Gimblett. She was wonderful, a professor of performance studies at New York University. Absolutely brilliant. As we traveled from Japan, to Bali, to Varanasi, to Nairobi, to Cairo, and to Jerusalem that academic year, we found that tourists were everywhere and local societies were producing performances for the tourists. We took the tourist performances seriously and began to study them.[5]

Another major influence was my collaboration with Victor Turner, beginning in 1979. He introduced me to Wilhelm Dilthey and to the anthropology of experience. The key idea here was to view culture not as a set of patterns, or rules, or behaviors, or traits, but as it was experienced by persons. The view is that culture is always in process, ever changing, and, as Victor Turner said, we all enter society in the middle.[6] So those were some of the books and persons that influenced my thinking.

Take us back to the very beginning. How did you get into anthropology? I was born in New York City in 1924. You're talking to a real New Yorker! When I was younger we lived in the West 70s, then a Jewish neighborhood. It was really a wonderfully diverse city. We would go to Rockaway Beach in the summer. My family had a room there. When we went, at Rockaway, there was a Jewish section, an Italian section, an Irish section, and an African-American section, and I questioned at the time, where is the melting pot?

The event that really changed my life in those early years was being accepted into Stuyvesant High School. Stuyvesant is a free public school, but you had to apply for it; it is one of the few that required a test for admission. It's based on meritocracy. It's actually harder to get in to Stuyvesant than Harvard. One recent year that I checked there were 20,000 applicants for 880 places. When I was there, more graduates went to Ivy League schools than at any other public high school.

I received a wonderful education and it was an amazing thing that all kids in New York could get that type of education. It was mainly known as a math and science school, and though I was actually in an advanced math class, my passion was in the humanities. I thrived at Stuyvesant. I read more good literature in those days than at any other time of my life. I would make lists of books, and I really read them. In the New York State Regents Exams I got a 98 in history. The New York State Regents are quite hard. I got a 96 in English, which pissed me off. I inquired about it and one teacher told me I made eight mistakes in spelling!

In those days, I had friends but I really was a lonely kid. I read a lot in my spare time, took walks, and worked part-time after school to make money to go to Broadway shows and sit alone on Saturday matinees way up in the balcony. But reading that great literature then, it led me to think of how great books were constructed. And during my high school years, 1938–1942, I published short stories and poems in Stuyvesant school publications. This, I believe, increased my sensitivity to stories and narrative.

But there was still tension in high school between the sciences and humanities. I decided to go into engineering, and I don't know why. I mean, I would've never conceived that I could've gone into academia. My father never went to high school and my mother went for a few years and then dropped out. And no one in the family had gone beyond that.

After World War II, I wanted to get away from New York and decided that I wanted to go "West." I had a New Yorker's conception of the United States. And I thought Ohio was West—wow, the real frontier. Wow! So I ended up at Ohio State, where I was from 1946 to 1950. After two years at Ohio State, I decided I didn't like the prospect of being an engineer. So, I went to the school's career advisory service and they did a battery of tests and did some interviews. They told me that I had the most aptitude for the social sciences, but I dismissed that because I couldn't conceive of how to do it.

Well, at Ohio State I met Elaine who's my soul mate, best friend, and life partner who has been with me for seventy years. She graduated a year ahead of me and was heading back to New York, and I wanted to be with her. So, I applied to Columbia engineering to finish my degree. And, I was surprised that I was accepted but they said that I had to take two summer courses because of differences in requirements. One was a calculus course, which was nothing for me, and the other was in the humanities. For my humanities course I selected an anthropology course by Alfred Kroeber, who was teaching at Columbia in the summer of 1947. I had no idea that he was probably the most distinguished anthropologist in the United States. He had just revised his monumental text, *Anthropology*,[7] so in class in '47 we went over that material. I was so excited and just flipped. I said to myself that this was for me, and I had to do this. Elaine, to keep me company—she was free that summer—she took an anthropology course with Ralph Linton. I read all of the books from Kroeber's course and most of the books from Linton's course in three or four weeks. And little did I realize that a mere seven years later, in 1954, I would replace Linton at Yale. He had passed away. I became a Yale assistant professor, and Kroeber eventually became my colleague because he was invited to teach for one year at Yale.

What did you like about anthropology?

Well, when I was young, ten or eleven years old, I read a book by Richard Halliburton called *The Royal Road to Romance*, and it's about a boy that traveled all over the world, sometimes living for short periods with indigenous peoples. I read almost all of his books and I was so enthralled with them. It was the idea of living and studying other people and their cultures. It was adventure—sheer adventure—that attracted me so much to anthropology. Maybe anthropology was *my* "royal road to romance." We should note that I entered anthropology because I thought it was romantic and adventuresome, not like today's entering students who want to study social justice and inequality to make the world a better place.

But I didn't really know how to become an anthropologist. So I went to the offices of Alfred Kroeber, where I told him that I wanted to do anthropology and I asked if one could make a living out of it [chuckles]. I also spoke to Ruth Benedict and Ralph Linton. They spent time with me and encouraged me. And it was very good. Then Elaine and I went for the weekend to Rockaway Beach to talk it over. I told Elaine that the salary may not be that great but she said go for it. So after that we told our parents—and Elaine's family was something! They invited us up to Westchester to discuss this issue; this is still 1947. And I get in their house and I find a room filled with aunts and uncles and everybody in the extended family. And the first question that they asked me was, "What is anthropology?" but I really didn't know because I had only read a couple of books and attended a few lectures. So, I said you travel around, study other people—and they asked me, "How much money do you make?" That was the real issue for them, if I could support Elaine. I didn't know what the salaries were but then anthropologists seemed to get along with it. I went through that torture for half an hour. There was Elaine's uncle, a Dr. Joe Rothenberg, who had asked no questions. He was the patriarch of the family. After about half an hour he got up and said, "Well, Ed wants to be an anthropologist," and then he left the room. And that was the end of the meeting [chuckles].

I went back to Ohio State so I could finish my undergraduate studies by taking all anthropology courses in my senior year. I worked under the leadership of John Bennett, Kurt Wolff, and Jim Schiller. Kurt Wolff was a sociologist who introduced me to the great figures in European scholarship, like Max Weber, Georg Simmel, Karl Marx, and others. And I learned basic anthropology from John Bennett. What I received at Ohio State was a warm, welcoming environment, which is what I needed at the time, so I decided to stay for a Master's. I became a teaching assistant, not to someone else's course but to my own introductory course, which I taught alone. After one year as a student I found myself the second year teaching anthropology. It was the best learning experience.

I recall one social psychology experiment that I learned about at Ohio State. You start with a big circle of persons in a ring. Then one person starts to tell a story that is whispered to the person to the left. That person would retell in a whisper the story to the person on his or her left. The process continued until the end of the circle. At that point they compared the original story to the ending one, and they were so different. Not just the details but the whole story. That is the key to everything that I ever did. I thought of stories and their retelling or of how culture is passed down from one generation to the next. Storytelling is constitutive, culture is constitutive, performance is constitutive. No performance is an exact duplicate of the previous performance, and that's how cultures change, every time in every situation. My work in narratives really stemmed from that idea.

This was reinforced when I went to graduate school at the University of Chicago from 1950 to 1954. I got a fellowship for $1,000, which I thought was incredible because at that time it was a lot of money. When I got there I found out that I had to pay $700 for tuition so only $300 was left! So Elaine went to work. We didn't even think about it being sexist that the woman sacrificed, although we did discuss it years later after feminism became prominent.

At Chicago, I studied with Fred Eggan, Robert Redfield, Sol Tax, Robert Braidwood, and Sherry Washburn. The system then was that Chicago had a basic graduate course in cultural anthropology, taught by Redfield; one in physical, taught by Washburn; and one in archaeology, taught by Braidwood. I was asked by Washburn after the first year to be his assistant in his physical anthropology course for the following year, and if I would've then I might have ended up studying primates in East Africa. But I didn't. I told him that I really wanted to be a cultural anthropologist and I didn't want to go into physical.

In the basic cultural anthropology course at Chicago, taught by Redfield, I was greatly influenced by George Herbert Mead's *Mind, Self, and Society* (1934), the founding text in symbolic interactionism. I also took a course with Herbert Blumer, a distinguished symbolic interactionist. After that first year Redfield asked me to teach his graduate course the following year. Not to be an assistant or to co-teach but to take over the course, using Redfield's syllabus. In other words, the first year I took the course, and the second year I taught it to incoming grad students. I had never heard of this before—I mean, a second-year graduate student teaching first-year graduate students at a place like Chicago! But having to go through this material twice was fantastic. I felt then that I was on the way to becoming a real anthropologist.

I did my dissertation on Native Americans. You have to remember that I started in anthropology in 1947, which was two years after the end of World War II. So, the work I was doing really goes back to the '40s and '30s. There weren't any paradigm shifts or major theoretical changes that occurred until much later. In 1948, I worked on the Navajo with Clyde Kluckhohn. We lived on a Zuni reservation. I got into the issues of culture change, acculturation, urbanization, and change in social organization.[8] There was a group of colleagues that coalesced around those theoretical and empirical interests. These were people like Ned Spicer and Evon Vogt and Ed Dozier, who was a Pueblo anthropologist at Chicago.

How did things evolve after you started teaching at the University of Illinois?

Other influences followed there. At Illinois I read the European poststructuralists and other European scholars. I participated in a faculty seminar beginning in 1980 that lasted for decades called the Unit for Criticism and Interpretive Theory.[9] There were some very great guys there like Andrew Ross, Cary Nelson, Larry Grossberg—all of whom have made names for themselves in cultural studies. I found that all of my postmodern readings, like reading Roland Barthes, were more humanistic. My interest in narrative was nurtured there. My reading of Bakhtin also influenced me.

In 1983–1984, when I was faculty director of the study abroad program with Barbara Kirshenblatt-Gimblett, we developed a deep interest in tourism. From Barbara I developed a deeper appreciation of the concept of performance and realized the extent to which performance was constitutive.

How did you get involved in the study abroad program? Was that what got you into the anthropology of tourism?

In 1982 I received a call from a Karl Jaeger in Bath, England, who was a philanthropist—an American. He had sponsored a study abroad program called the International Honors Program, which took around twenty to thirty undergraduate students from different American universities around the world, traveling by air. He said that Gregory Bateson had been director for one year, and he said, and I paraphrase, "I'll pay some of your salary, all of your airfare. I'll pay all of the airfare for your wife. She can be the dean of students. You can pick the itinerary. You can pick the courses that you want, and where to stay. You can select a second faculty member." I accepted over the phone. Jaeger actually arranged for the students to stay with local people in local homes for about six weeks in each locality. So, I figured, "Where would I like to visit that I have never visited?" I took the group to Kobe, Japan, for six weeks, then Bali, then we took a one-week vacation to Singapore. Then to Varanasi, by the Ganges, which was amazing. Then to Nairobi, where I studied Maasai tourism, and then to Cairo and Jerusalem. I was on the trip for the entire 1983–1984 academic year. Barbara Kirshenblatt-Gimblett and I worked well together, and have become lifelong friends.

How did you know her back then?

I didn't know her. I was introduced to her through some other people's recommendation. She and her husband came to Urbana to discuss the trip. Whenever we're in New York she and her husband and my wife and I go out for supper. Barbara and I love each other. I mean, we really grew close.

One of the requirements of the course was that the students had to write a research paper about each country. Well, we got to Bali. Bali has as many tourists as Balinese. So most of what the tourists see are tourist performances but there are also Balinese performances for Balinese. Tourists would go to all of the places that anthropologists go. It raised questions about the relationship between tourism and anthropology, and how similar and different they were.

It related to the whole issue of authenticity. We went to a place called Mayers Ranch near Nairobi, run by ex-colonial British people. It was a Maasai tourist performance.[10] And someone would say that this is not really the Maasai; it's just made up for tourists. They denigrated tourists in those days. They said if you really wanted to know about the Maasai you don't go to a tourist performance, you go out to the Maasai reserve, which is twenty miles away or forty miles away. And you get there and they say, "You should've been here twenty years ago!" [laughs]. And then you can look back to when the Maasai migrated to Kenya. And you ask, which is the authentic Maasai performance? The question of what's authentic becomes meaningless in a situation like this. We learned very early, Barbara Kirshenblatt-Gimblett and

I, that authenticity is a red herring. Authenticity makes you compare what you currently see with something that is a fantasy.

We were amused in Bali meeting a woman from the *Village Voice* who came looking for the authentic Balinese. She was a reporter doing a story. She had known Barbara from New York and she made it clear to us that she was looking for the real, the authentic Balinese. So I said, "Go search!" I mean *all* of what Balinese do is authentic, as it's all real Balinese.

I recall a scene in Woody Allen's 2011 movie *Midnight in Paris*. It's a movie about time travel, and the hero meets this girl in contemporary times, and they get transported back to the 1920s. The hero feels that the 1920s was the Golden Age. Then they get transported back to 1890 and the girl says that that was the real Golden Age. The hero says that the people in the 1890s thought that the Golden Age was the Renaissance, and if you keep searching you can go back to Kublai Khan. If you're looking for the authentic Balinese you always have to go to another place or another time. What's authentic is always located elsewhere. The basic problem with a lot of the tourism work at the time was that to the extent that researchers believed in authenticity, they didn't see what was happening in front of them. They didn't take Balinese dance performances seriously. Constructing culture for tourists represents a great innovation and a great subject for anthropology, and it's happening all over the world.

There are other implications. I later studied a reconstructed village called New Salem in Illinois where Abraham Lincoln lived in the 1830s. It's a huge tourism site that actually influenced academic scholarship on Abraham Lincoln, as it was so prominent and because it emphasized the early years of Lincoln's life.[11] On the study abroad program, the students began writing their research papers on tourist performances since they were so accessible.

I believe that tourism influences academic scholarship, and anthropology influences tourism. In Bali in the 1930s a group of Western intellectuals were so taken with trance and the Barong figure—and wrote about it—that the Balinese, sensing this, made it more prominent in their culture.

But if the question of authenticity is meaningless, do you see any value in employing the term?

I see value in the term *authenticity* when the people use the term *authenticity*. Because then it's like any term and you want to find out the meaning of it.

So you are approaching it as a folk category?

Yes, like a folk category. And you can spend a lot of time trying to figure out what the people mean by it or what the tour guide or tour companies mean by it.

Dean MacCannell, in *The Tourist* (1976), used the concept of "staged authenticity," and he has subsequently written more about it in a recent paper with a title like "It's not about authenticity at all."[12] I believe him.

Authenticity has not been a key concept for Dean, just as it has never been a key concept for me. I never initiated research to study authenticity. But as an ethnographer, when I study a site like New Salem, which called itself an "authentic reproduction," which seems like an oxymoron, I have to determine what the producers of the site mean by the term.

In the experiences with luxury tourists that served as the basis of *Culture on Tour* (2005a), I had this initial conflict with the president of the tour company. She had a conception of how to present the cultures in Indonesia. I said to her that one of the things in Bali was a Frog Dance that was invented for tourists in 1971. I arranged a performance of it for the tour group. It was a lovely dance. And she asked, "What was it before it was invented?" [chuckles]. It disturbed her whole narrative of the tour being authentic, which was then accepted by tour agencies. And the fact that it had a local history—that tourism events have a local history—was not considered.

How did you relate that to anthropological work already done on tourism?
I didn't start out by reading the tourism literature. As I mentioned before, there have been different lines of development here in my thinking. One of them has been the theory of constructivism—where things are constituted—and the other one is the experience of seeing tourism and anthropology studying the same types of objects albeit in completely different ways. It was the era when I learned about postmodernism and interpretive theory.

I never had a course on tourism. I never had a mentor in tourism. I just developed this interest and began to study tourism. At about the same time, as I mentioned, I developed an interest in interpretive theory and postmodernism, and I wanted to apply that theory to tourism. Later I connected with tourism scholars—Dean MacCannell, Nelson Graburn, Erik Cohen, and others—very thoughtful people. I connected with these people after I had done most of my studies, so I really wasn't influenced by them.[13] So I sort of backed up into the anthropology of tourism. I didn't have the advantage—I mean it would've been great if I had been a student of Nelson Graburn's going to Berkeley.[14]

You also mentioned Victor Turner as one of your influences. He had a different relationship to tourism, because he wrote on pilgrimage and ritual, and his work on that has made its way into the anthropology of tourism literature. How did he influence your work?
Victor Turner was not really into tourism, but of course he understood everything I said. We would have long, intense conservations over the telephone. These were the years from 1979 until he died in 1983. But theoretically I learned so much from him, and also he acted as a kind of mentor to me.

Victor and I were the same generation. I graduated and got my PhD in 1954, and he got his degree from Manchester in '55. We seemed to be at the same meetings and conferences and got to know each other. Anthropology was a lot smaller then and the annual meetings much smaller. We had similar interests but it wasn't until we both went to a conference on narrative in Chicago that we really began to work together.

This was the period in which I reinvigorated my anthropological self. I felt I'd reinvented myself in the 1980s with postmodernism and work on tourism. Vic's main goal was to fight structuralism as too formal, and he wanted to introduce human beings into the equation.

It was at a conference in Chicago on narrative, in 1979 I believe, that Vic, Barbara Myerhoff, and I, the three of us, developed the idea of a symposium at the next AAA annual meeting on the anthropology of experience. I spent a lot of time with Vic and with Barbara at that conference. The basic idea was that culture does not consist of just patterns or traits or ways of behaving. Culture for us was what it means to people, how people actually experience their culture. Which is a very humanistic approach.

So we had the AAA symposium in 1980 on the anthropology of experience that was published as an edited volume and is still in print.[15] Vic was a great man and a great anthropologist. Barbara Myerhoff had passed away, and Vic wanted me to write the introduction to the book. I wasn't sure I was up to the task, and I said so to Vic. Then he would boost my ego, and he would say that some of my papers were as good as any others he had ever read. He'd bump me up. He was a really wonderful mentoring influence.

Vic died while we were in Varanasi. Richard Schechner communicated with Barbara Kirshenblatt-Gimblett and told us that Victor had passed away. I carried through with *The Anthropology of Experience* as editor, but I placed Victor's name as the first author. It was, after all, his idea. But Vic had not been as familiar with the current postmodern literature as I was; a lot of the thinking about narrative was new to him.

My collaboration with Turner was a turning point for me. Not only intellectually but as part of a cohort of anthropologists working on similar issues. I got involved and accepted into his whole network of people. The contributors to *The Anthropology of Experience* and to *Text, Play, and Story* (E. Bruner 1984b), such as Renato Rosaldo, Barbara Babcock, Jim Peacock, Jim Fernandez, Jim Boon, and others were struggling with the same postmodern issues.[16]

After Victor passed away in 1983, there was a series of panels, possibly four or five, at the AAA annual meetings that the anthropologists Kirin Narayan and Smadar Lavie organized, as graduate students, with Renato and a group of people who were studying the same kind of issues.[17] When I invited people to participate in the book that became *Text, Play, and Story*, Renato told me that there was a guy out on the West Coast that I should really invite. I didn't know the person. I said "What's his name?" "Jim Clifford," he replied. I should really have invited him. But we had a full lineup and I didn't, which was one of my big mistakes.

I then participated in the 1980s AAA panels, and one in 1984 or '85 with Jim Clifford and the *Writing Culture* people.[18] We used humanistic, postmodern approaches. I recall that in all of these symposia they asked me to be a discussant. I would have liked to present a paper but the organizers never asked for that. So I was a discussant of a whole series of panels. There was one paper that Paul Stoller presented. I admire Stoller's work very much. He gave a paper on his relationship to his informant, and I thought it was a little overdone, so I made some comments about it.[19] When I was president of the Society for Humanistic Anthropology (1989–1991), he was on the board, and we were active together in that.

You certainly were the "restless anthropologist"!
I think one of the main points here is that as I moved from American Indians, to Indonesia, to postmodernism, to tourism, it was all with and through different colleagues. With each of these transitions in my work, I worked with new people, and I seemed to have moved into the group that was working in each new area. That was the whole point of the "Restless Anthropologist," my (2012) paper on moving from one place to another. With each shift, I was starting over, but I was fortunate that I was accepted by each group on the basis of what I did. As I changed areas, I joined the anthropologists who were already working on the same issues.

Could it be that you were able to innovate, to be free to develop your own way of thinking without being bogged down, because you were coming across these theories independently?
Yes, that's absolutely correct. I came into these theories without having a backlog of previous study. And I found some amazing people whom I learned from. But despite moving through different cohorts, there was also a continuity, a theoretical thread that went all through my work from the very beginning—constructivism as created through narrative and memory. Erik Cohen, a sociologist and a tremendous scholar of tourism, once wrote to me that my postmodern studies of tourist productions were teaching everybody else how to do tourism studies. It was so flattering and uplifting.

What do you think Cohen meant by that?
My approach to tourism. I studied tourist productions. By this I mean a dance drama, a craft village, a reconstructed village or community. Whatever was put on display for tourists. I regarded those as productions, and I took them seriously. I studied them as an anthropologist would study ritual. I focused on the production. Who put on the event, who produced it, what was it for, who was the audience, what was the audience reaction to it. There were a lot of different and excellent approaches to tourism and a lot of different ways of

doing it. My way of doing it was to take tourist production as a performance and as a ritual and to study it in all its dimensions and its multiple layers of meaning. I tried to analyze the production from the beginning to the end, and how it moved through time and space.

But you also took the position of a tourism producer at times—as a tour guide or study leader.
When I went with tourists, it was mainly as a tour guide—which I think is a fine way to enter the tour group. I stayed with them for the entire trip, listened to their conversations, observed their reactions to various sites, and answered the questions they asked me, as tour guide. Becoming a tour guide, I believe, is one of the best routes to get yourself embedded in the tour group.

You've written about how taking that role also gave you insight into the dynamic between tour operators and local producers. Tour operators aren't always the nefarious entity trying to dupe the traveler into "staged authenticity." Often they think tourist productions are "real" too.
Yes, but the tour operators usually want to organize their tour in terms of the imaginary, the pretour image, because that's what the tourists expect. I was right there as it was being negotiated in the touristic borderzone.

Can you talk more about the borderzone, since it's become such a hugely influential concept in the tourism literature?[20]
The way I conceive of the borderzone is that typically the tour group forms in what is usually a Western country (or it could be any large city), and they have a tour guide, and they start out from there on tour. While on tour they go back to their hotels in the evening for supper, and then they depart from the hotels in the morning to tour Bali or Tuscany or wherever. The tourees—the ones that have been toured—in the evenings go back to their villages, and they go back to their lives. But there are certain places where tourists and those toured meet, and that is what I call the touristic borderzone. In the borderzone, tourees want to meet the expectations of the tourists. It's a space of encounter, although I found that tourists interact with other tourists as much as they're interacting with the people that they tour. But the borderzone is where they interact, and that's the place where conceptions can be changed or reinforced.

The borderzone is obviously a liminal space; it's temporary, "betwixt and between," to use Victor Turner's phrase. Do you see it as being a creative space, too, in the sense that Turner wrote about liminality?
I disagree with Turner's concept of liminality. He was my mentor, and he led to my reinvigorated self at a certain point in my anthropological life. But I criticized the concept of liminality to him. His concept is that basically you're

living a normal social life and then go into this liminal space, and this liminal space is one of creativity and *communitas.*[21] But then after it's over, you return to normal social life—in his conception, A. R. Radcliffe-Brown's space. So I told Victor that he's really a structuralist after all! And he didn't like that. He told me his main goal was to contest British structuralism and Claude Lévi-Strauss's structuralism. He wanted to humanize the subjects. We talked about it. But my point is that in normal social life there are times that are departures from the norm, but not necessarily liminal. People don't just live in fixed slots, in roles; they construct their lives, and Turner has made normal social life a very dull place because you go there before and after liminality and *communitas*—that's what the old British ethnographers wrote about. Vic said that he was writing against Radcliffe-Brown and E. E. Evans-Pritchard and Meyer Fortes and other well-known British social anthropologists who had a rather fixed conception of a society, and he wanted to show that there was a part of social structure that was disrupted in these liminal events and that they hadn't taken account of those. That was his main point, and I could understand his perspective.

In your writing, you've approached the borderzone as a place of encounters between imaginaries and experience. How do you conceptualize the relationship between them?

I have a fourfold conceptualization. First is the imaginary. The second is the experience. The third is the reality. And the fourth is the telling, which may replicate or depart from the imaginary. So the fourth one and the first one essentially are about the creation and replication of ideas about tourism.

The imaginary consists of the expectations that the tourists have before they go. There are no naive tourists. They have a conception of what Cambodia will be like or Angkor Wat will be like or some view of what the Maasai will be like based on general Western beliefs in the culture and media. They formulate the imaginary through schools, through television, through all kinds of media in their own culture before they go on a trip. Then they go on the trip and they have the experience, how they actually experience the trip. Now, there is a really interesting interplay between the imaginary and the experience because some people will conceive of the experience in terms of the imaginary so that they say, "Oh it's just as I imagined it to be!" and other people will have an experience which doesn't accord with the imaginary. And I think this is a point which applies to ethnography—indeed, it's the key to ethnography—because we go to a new culture and we've read the literature and we have a view of what's there, and then something unexpected happens or the people say something that doesn't accord with what we had expected.

The astute ethnographer will take that as a point of entry and analyze that, and it's how we learn something new. If the ethnographer goes to a people and everything is just like he imagined it to be before he went, then we wouldn't learn anything new. The new things emerge in the discrepancies. So that's the imaginary and the experience. Now the reality is what actually happens but nobody knows what that is. There's no way of getting at that. It's an ontological reality, and you don't get at it unless you talk about it and when you talk about it, you're already telling it. So, the reality is there, but it's an unopened chapter.

So we have the tourist's expectations, based on imaginaries, and then the actual experiencing of the place. When these two things come together, they may clash, and one makes meaning of the experience by reverting back to the preconceived notions, the imaginaries?
For some tourists the imaginary and the experience are similar, and for other tourists there are discrepancies.

Is the expectations gap—the lining up of the imaginaries and the experience, or not—is that reality?
Reality is unknowable. There isn't any such person except God who knows all. So we use the concept of reality but it is an empty category. It's empty.

So nobody can actually know reality because everybody is going to have their own perspective. What is the role of the telling in that?
The telling of the tourist experience is that they talk about what happened. On the tour at dinners or on the bus, they talk about their experiences to fellow tourists, but usually they talk about the tour after they return home. They tell others about it. Now sometimes, the way of the telling depends on the audience. You know, I use the example of a student who comes to the university for the first time and he finds it a very confusing place with lots of big buildings and places to go and lectures that he's not heard before and dormitory life that he's got to get used to. He says to his roommate, "This place sucks," but then his mother calls him up and asks, "How are you doing in college?" and he replies, "Oh, it's great, mom." You know? I mean, the telling varies with the audience. And the telling varies over time. A very important thing which hasn't been done enough in tourism is an analysis of post-tour tellings. They change over time, and often they are with a set of pictures or a video now. And what you remember of the tour and how you talk about it does change. I wish somebody would do a long-term study of that and find out how that would happen.

Isn't telling also creating new imaginaries, especially for the audience?
That's a very good point, because as you do the tellings—particularly the post-tour tellings— you are creating new imaginaries. This is storytelling, or

the construction of a narrative. I'm very interested in the whole thing of the post-tour narratives to understand how you go on a trip and to whom do you tell the story of the trip. How often do you tell it? How does it change over time? You go on the trip, and you come back to your family, and you tell them all about it, but how often or how long do you tell them about the trip? And when does it come up? Maybe when other people are planning trips. I mean, you run on repeat after a while, and you run out of people to tell it to.

Doesn't this go beyond even the act of recounting the story of a trip? Don't these stories inform the meanings that the teller subsequently makes of other destinations, even if it's a totally different place?
Sure, sure. They're blurred. They're blurred concepts. And this is what I mean by doing a processual analysis. I think of processual analysis as being more dynamic and not as a fixed thing. All tellings are constitutive. The role of a tourist, for example, is not a fixed category that a person just occupies and then you're a tourist. You construct the role. It's always moving.

Now, what's the difference between a narrative and a metanarrative—a term you and Barbara Kirshenblatt-Gimblett both use in relation to tourism?
A metanarrative is a narrative about narratives. It's a metacommentary. For example, we have a story about Bali as a South Seas island paradise, with dance dramas, beautiful women—altogether a very picturesque place. There is a similar narrative about Hawaii. Both are Western constructions. If one were to analyze the processes of construction in both Bali and Hawaii, and in similar islands, that would be a metanarrative. It's like a metacommentary on the narratives you have about particular places. This is how I use the term although I realize there are other uses, and scholars like Jean-François Lyotard (1979) have criticized metanarratives.

When I was a tour guide in Indonesia, as I describe in *Culture on Tour* (2005a) and earlier in our conversation, the head of the tour agency accompanied us. We had different and conflicting conceptions of the tour and of tourism generally. She saw tourist performances as representations of an authentic culture, whereas I saw the performances as staged, constructed for the tourists, and historically situated. Hers was the standard narrative in the tourism industry. Mine was a postmodern perspective. I call these two metanarratives as they are grand conceptions about tourism generally. They apply not to a particular place but to tourism generally. The tour owner was greatly upset about my deconstructing her narrative, and she eventually fired me. I think she felt I was ruining her tour. I wanted the tourists to have more of an analytic anthropological perspective but I failed in that. They sided with her, not me, and I was isolated during periods as a tour guide. They had the same imaginary and conception of tourism that she had. I was trying to burst that and I was trying to reveal that so much of what tourism is shown in Bali is not the old Balinese dances and dance dramas but ones that have been modified for tourists.

I thought that my approach was better than what the regular tour was. I was deconstructing the tour as I was doing it, and I really think that it was too analytical or anthropological. I mean, they didn't come on tour to be in a classroom for education. They really didn't want an academic lecturer. They wanted me to tell the stories that they already knew but just in more depth and more nuanced. And then there was one tour group that I led by myself, where I got very familiar with those tourists. I gave a lecture on authenticity, but they replied that they were not concerned with authenticity or some reconstructed past. They just wanted a good show. Nonetheless I think that they were a level above the naive tourist who just goes along with what's being presented. When I'd point out things that were being constructed for them, they didn't care. It was nice. I mean, they enjoyed it.

Some would say that anthropology is also essentially about constructing narratives.

Absolutely. That's what an anthropologist does and that's what a tourist does. I mean, you have the experiences on the tour and you usually don't relate them until after the experience is over. The same for ethnographers; we have experiences but don't convey them until after the fieldwork is over. That's when you start to write, and writing is its own creative, constitutive process. And you see things that you hadn't realized before as you're writing. Writing is a very creative process. You start writing and you let your mind wander and improvise, and you pick up various things. Tourists are doing the same thing in their post-tour tellings. They're making up—I mean, they're telling stories based upon the conditions that they're in, the audience, where they are in life. So, the idea of a processual analysis is more that you do something not as the old anthropology did in America and Britain in the 1930s, '40s, or early '50s. You move away from that positivism, and you think of process and change and dialectic and interactions, and I think that's what I mean by processual.

You say that tourists are telling stories about their travels based on the specifics of their lives. Do you think they have ethical concerns that shape their stories, too?

They do have ethical concerns. I tell the story of once on a very upscale tour in Africa, we were in a group of Land Rovers going through a poor village and a group of children dressed in rags ran after the autos. Very poor children. And the tourists talked about that and remembered it. The event brought to mind the discrepancy between their wealth and the poverty of the people in Africa.

That was a concern to these upscale tourists. The tourists were good people, you know? Their wealth was so much more than the wealth of the Africans. It disturbed them. I think there are more ethical kinds of considerations.

Speaking of ethics, what about your method of studying tourists while ostensibly being one yourself? Did your fellow travelers know that you were going to write about them?
I usually said that I'm an anthropologist studying tourism but they never really picked up on that or withheld things from me. I don't think I told them that on every tour. When it came up I'd tell them. I suppose when I started being a tour guide and started writing about tourists, I didn't want to be so much of an ethnographer that I'd go to the tourists and ask them to tell me their stories. That would make my anthropological role too obvious. But maybe there are other ways to collect that data.

I have this funny story. One of the things that I wrote about was the tourists were taking pictures of the Indonesians and I started taking pictures of the tourists taking pictures. The president of the tour company said, "It makes our tourists feel uncomfortable. Stop it." So I thought, what about the Indonesians? What do the Indonesians feel about tourists taking pictures of them without asking permission? Anyway, there was too much of a break between her conception of what a tour guide is and what I wanted to do. I mean, I was there to do anthropological fieldwork. And she wouldn't let me do it. I understood her goals, and that she had a business to run.

So you were the ethnographer taking pictures of the tourists, and the tourists were taking pictures of the Indonesians. Could this be why anthropologists were slow to take tourism seriously—because there's too much similarity between the behaviors of the ethnographer and the tourist?
It's similar and different. Bad ethnography is called touristic. Obviously what tourists do is very different; I mean, they stay for much shorter periods of time. Anthropologists in the past regarded tourism as too commercial and tacky. I'm talking about guided tours rather than backpackers. Tourism was not part of the aboriginal culture they were reconstructing. They left it out of their monographs and reports.

Don't anthropologists do that, too?
Yes [chuckle]. Because what anthropologists do is go to these strange places and tell stories about it. It's not an exact replica of what they saw. It's what they construct as a story, and what they construct as a story is in part determined by the conventions of what is ethnography. The anthropological conventions structure in a way how anthropologists see things and how they write about them. Anthropologists have always wanted to differentiate them-

Figure 10.1. Edward Bruner photographing Maasai dancers at a tourist performance, 1995.
Photo credit: Edward M. Bruner.

selves from others who were around them in these exotic places, like travel writers and explorers and colonial administrators and tourists. I don't know why tourism was left out so much, because everywhere I've been traveling in the world there have been tourists.

I had a momentous event, which I've written about,[22] of taking a group of tourists in Bali to see to see a temple festival, and there was Hilly Geertz doing ethnography. Tourism and ethnography were competing in the same space. She was at a Balinese temple performance when I bumped into her with the group of tourists. I had known Hilly and Cliff for many years. And she said, "Ed, come to visit me after the tourists leave, but I don't want to talk with the tourists." She had other ethnographic things to do. Then about five minutes later, she came over to where we were sitting and she said in effect, "I'd like to take the tourists to meet a painter named Togog." Hilly studied Balinese paintings, and she thought the tourists might buy one of his paintings. So we all trooped off together. It just impressed me so much that here I was a tour guide being led by Hilly, who was then a master tour guide and master ethnographer, to visit an informant named Togog. And Togog, that artist, had been an assistant to Gregory Bateson years before. And in a way you can say that Togog was Bateson's assistant, but Bateson was Togog's tourist. And this blurring of the lines of what's authentic and what's real and what's tourism and what's ethnography made you reflect and think about it.

Didn't many anthropologists of that era totally overlook the fact that there were tourists running about their fieldsites?

Yes, but if you look earlier, at the anthropology of the 1910s and 1920s, the Boasian and Kroeber kind of anthropology, and British social anthropology, they completely overlooked the political structure. I mean, Robert Lowie (1935) wrote a book about the Crow. But he didn't mention that the Crow were under the control of the American Bureau of Indian Affairs, which was part of the federal government. He was trying to reconstruct the quote-unquote "old culture." And he included buffalo hunts in his book—which the Crow no longer did when he was there. He put that in but he left out the whole thing about how the Crow were embedded politically in the American system.

So tourism was treated the same way. It was just left out. It was not considered a legitimate kind of thing. And that's part of what I mean about taking tourism seriously. You have to study what's there in front of you. You can't be arbitrary and in a sense leave things out and have prior conceptions dictate what you see. Recently, the inclusion of tourism has just blossomed and mushroomed. It's become an accepted part of anthropology now.

Why do you think that shift occurred? It often gets linked to the postmodern movement associated with the volume *Writing Culture* (Clifford and Marcus 1986). Would you agree? Do you think the collaborative work you mentioned earlier, your work with people like Victor Turner, had a role?

In 1984 I wrote the introduction to *Text, Play, and Story*, and I called it "The Opening Up of Anthropology."[23] I said that what this newer anthropology is about, is that it greatly increases the scope of subject matter of what you can legitimately study. And it was not restricted to the old idea of what an ethnography should be. We didn't all have to do *We, the Tikopia.*[24] About 1980 or so, anthropology had opened up and blossomed, and anthropologists could do all kinds of things that they never thought of doing before. We expanded what were considered legitimate anthropological topics.

Then we had the whole phenomenon of globalization and mass travel. All of those things were part of the newer anthropology of tourism. I think that the structure of tourism is built on entrepreneurs who established tour agencies; tour companies who decide on itineraries. Now in every place I've been, like in Indonesia, there is certainly a local director of tourism, a government tourism agency. It was the Indonesian government that decided to build infrastructure in Bali to facilitate and develop tourism. The government bureaucrats and business people are very involved with the development of tourism. They have the power. So there is an exploitative aspect of tourism. The profits don't go to the people although some Maasai and Balinese are now producing themselves for tourists. They're not using white intermediaries, and I think many indigenous groups in the future will be more and more producing themselves.

"Indigenous tourism," where the tourism experience is fully designed and implemented by locals, is a growing niche.[25] Do you see this as viable—or at least more viable than the old "community-based tourism," in which outsiders worked together with locals?
Well, one of the things that I've observed in a number of groups is that external tour operators are able to bring in more tourists. Indigenous people, in terms of social justice, could produce themselves and get the profits from it: find out about the money trail, where the money goes, how much money goes to indigenous people who put on these performances. Dennis O'Rourke really talked about that social justice dimension in his film *Cannibal Tours.*[26] To me, the people who are being toured should get more of the profits, and they should be producing themselves. But the problem is, say in Kenya, who has access to tourists? It's the tour agents—Thomas Cook and Abercrombie and Kent. They're the ones that have the contacts. I looked into this at Abercrombie and Kent. Their head of tour groups in the Maasai Mara reserve said, "We have five villages arranged that we can go to." So if one local group says, "We want more money," they can just cut them out and go to another village.

Looking back on your tourism research, do you have any reflections on your methodology?
There are very many different approaches to tourism, and I wouldn't tell someone to do what I do, or say my way is any better than anybody else's. If I could do it again, I would like to do what Julia Harrison did in studying tourists who had returned home.[27] I would pay more attention to post-tour commentaries. I think there are so many different ways of approaching tourism. I wanted to apply a theory and a methodology to the study of tourism.

Are there any other things you wish you could have done?
Yes. I could've written more if I had been bolder. When I was younger I could've explicitly criticized some of the work of Lévi-Strauss, Edmund Leach, and Rodney Needham on asymmetrical cross-cousin marriage, and applied my new theories to them. I certainly could've done that. I could have put some things together and written more books; when I got to know Victor Turner, I realized that he published everything twice. He would publish in a journal or book chapter and then he'd put them together in his own book. But

academics often feel that they have not done enough. Still, I've been on the cutting edge of many things that I have done. I've been accepted by people who are out front in different areas. So that's my good feeling about myself.

I'm glad you mentioned that. We started out this interview with a question about your legacy, but I don't think you ever answered it.
Well, we talked about my role, how I conceive myself in the development of the anthropology of tourism, and how I see myself as an anthropologist. I didn't say what I feel about myself. I'm a modest guy and a little self-deprecating, and I certainly don't want to be boastful. But I think I've been very innovative in what I've done throughout my life, through my early work on constructivism and kinship and using symbolic interactionism. I was very innovative in studying urbanization in Sumatra, and certainly in the anthropology of experience. I was an early adopter of postmodernism in tourism, using critical theory before many other people did.

So, there's a side of me that's very pleased with what I've done in anthropology, and it's certainly been recognized in the profession at large. It really does make me feel very good. I see some of my colleagues who, as they get older, they sort of fade out. I wrote a paper on "The Aging Anthropologist"[28] in which I refused to accept the definition of an old person in American culture. I constructed for myself my role as an old person. You have to fight stereotypes of older Americans and not just accept them.

So continuously thinking about and writing on new topics keeps you young?
Not really young—after all, as we are speaking I am ninety-one.[29] But I exercise most days, swim or walk or go to physical therapy, and I watch my diet. I've had a long and blessed life and I can still drive and exercise, I can still write, and I dress sharp. I have a wife who loves me and children who adore me. Things are pretty good.

NOTES

1. Leite had organized an AAA retrospective session on Bruner's work and developed a proposal for a volume arising from it, one that would include an edited "conversation." Inspired by Fred Myers and Barbara Kirshenblatt-Gimblett's (2001) published interview with Annette Weiner, she intended to provide a window into the thoughts and experiences that influenced the development of Bruner's approach to tourism, as well as his perspective on major theoretical issues in its study. Like the Weiner interview, this one has been edited and synthesized—with Bruner having been granted the final word in what was included.

2. Di Giovine also researched and wrote the endnotes, and his student, Aaron Gallant, tracked down and prepared the references. The three original interviews

were transcribed in full by Liselle Milazzo, Tayler Ulmer, and Nicholas Ogris, all then graduate students in the Department of Anthropology at SOAS, University of London.

3. Graburn (2005); see also Graburn and Leite, this volume. According to Graburn (personal communication, August 26, 2017), the informal remarks were based on his early reading of Bruner (1955; 1956b) in graduate school.

4. See, for example, "The centripetal forces of the life of language, embodied in a 'unitary language,' operate in the midst of heteroglossia. At any given moment of its evolution, language is stratified not only into linguistic dialects in the strict sense of the word, . . . but also—and for us this is the essential point—into languages that are socio-ideological: languages of social groups, 'professional' and 'generic' languages, languages of generations and so forth. From this point of view, literary language itself is only one of these heteroglot languages—and in its turn is also stratified into languages. . . . And this stratification and heteroglossia, once realized, is not only a static invariant of linguistic life, but also what insures its dynamics: stratification and heteroglossia widen and deepen as long as language is alive and developing" (Bakhtin 1981, 668).

5. Barbara Kirshenblatt-Gimblett (1998) would also publish her own study on tourism and heritage, based in part on these experiences.

6. Citing their volume, *The Anthropology of Experience* (Turner and Bruner 1986), this quote is mentioned in Bruner (1994, 407).

7. Kroeber (1948).

8. See Graburn and Leite, this volume.

9. The Unit for Criticism and Interpretive Theory began in 1977. See Bruner (1986a) for discussion of this interdisciplinary group, its participants, and the material explored.

10. See Bruner (2001); Bruner and Kirshenblatt-Gimblett (1994).

11. This research informed Bruner's (1994) problematization of authenticity and certain forms of postmodernism. See also Little, this volume.

12. "Why It Never Really Was About Authenticity" (MacCannell 2008).

13. Here Bruner is referencing three of the early scholars who in the 1970s paved the way to legitimizing the anthropological study of tourism: sociologist Erik Cohen, who beginning in the 1970s wrote a series of influential programmatic and theoretical articles, including "Toward a Sociology of International Tourism" (1972) and "A Phenomenology of Tourist Experiences" (1979a; see also Cohen 2005); sociologist Dean MacCannell, who first wrote on staged authenticity in tourism in 1973, a topic elaborated in his acclaimed book, *The Tourist: A New Theory of the Leisure Class* (1976); and anthropologist Nelson Graburn, who organized a series of foundational symposia on tourism in the early 1970s, edited the pathbreaking *Ethnic and Tourist Arts* (1976), and contributed "Tourism: The Sacred Journey" to Valene Smith's *Hosts and Guests* (1977). Graburn also developed the world's first undergraduate Anthropology of Tourism course (Graburn 1980). On the emergence of anthropological and sociological studies in tourism in the 1970s and beyond, see Nash (2007).

14. Ironically, Graburn says his graduate studies were informed by Bruner's work. "Of course, I was very familiar with Bruner's work from [my graduate studies

at] U Chicago and used it as inspiration for my 1963 dissertation. I am absolutely certain the two of Ed's papers from 1955 and 1956[b] were what I was citing [in the 2005 festschrift comments (Graburn 2005)] and which I used before" (personal communication, August 27, 2017).

15. See Turner and Bruner (1986).

16. See Bruner (1984b).

17. In the 1980s, Bruner was involved in a number of panels organized by Lavie, Narayan, and Rosaldo for annual meetings of the American Anthropological Association. He was a discussant for their 1984 two-part panel, "Persuasions and Performances: The Poetics of Self and Society," chaired by Edith Turner and Rosaldo, and again in their 1988 two-part session, "Unveiling Agendas: Person and Power in the Creation of Anthropological Knowledge," chaired by Rosaldo. Lavie, Narayan, and Rosaldo also organized a 1986 two-part panel featuring many of the same speakers (Edith Turner, Paul Stoller, Johannes Fabian, James Fernandez, etc.), but not Bruner, titled "'Othering': Representations and Realities." However, it is likely that Bruner attended the panel. In addition, at the 1985 American Anthropological Association annual meeting Bruner presented the paper "Tourist Performances and the Ethnographic Original" with coauthor Kirshenblatt-Gimblett in a two-part invited session organized by Gelya Frank and Schechner, titled "In and Out of Boundaries: Authenticity and Inauthenticity in Cultural Products, Performances, and Descriptions." While Lavie and Narayan did not organize the panel, it featured similar panelists, including Paul Stoller, E. M. Broner, and Richard Schechner. My thanks to Ushma J. Suvarnakar, Director of Meetings and Conferences at the American Anthropological Association, for locating these panels in the AAA archives.

18. See Clifford and Marcus (1986).

19. This occurred at the 1988 panel "Unveiling Agendas: Person and Power in the Creation of Anthropological Knowledge," for which Bruner was discussant. Stoller's paper was "A Dialogue on Anthropology between a Songhay and an Inquirer." According to Stoller (personal communication, August 1, 2017), this paper became part of his ethnography in honor of his mentor, the possession priest (*zima*) Adamu Jenitongo (Stoller and Olkes 1987).

20. For more on Bruner's concept of the borderzone, see Bruner and Kirshenblatt-Gimblett (1994), E. Bruner (1996b), and chapters by Leite et al. and Simoni, both in this volume.

21. For discussion of Turner's idea of liminality and *communitas* within the realm of tourism and pilgrimage, as well as critiques of the theory, see Di Giovine (2011).

22. This story is recounted in E. Bruner (1996b). Bruner is referring to Hildred Geertz and Ida Bagus Madé Togog. Geertz and Togog would coauthor *Tales from a Charmed Life: A Balinese Painter Reminisces* (2005), the third in Geertz's trilogy of monographs on Balinese culture, art, and religion.

23. Bruner (1984a).

24. Firth (1937).

25. See, for example, Bunten and Graburn (2018).

26. *Cannibal Tours* (1988), 70 min., Ultimo (New South Wales: Screen Australia). Bruner published a review of the film in *Cultural Anthropology* (Bruner 1989b).

27. Bruner is referring to Julia Harrison's 2003 ethnography, *Being a Tourist*, which is frequently assigned in anthropology of tourism classes. See Harrison, this volume. A recent ethnography in which the anthropologist studies tourists before, during, and after their trip is Hillary Kaell's *Walking Where Jesus Walked* (2014), which examines Christian pilgrimage tourism to the Holy Land.

28. Bruner 2014. For a discussion of Bruner's changing of field sites—which was also given as a paper at an earlier AAA conference—see Bruner (2012).

29. The interview was completed in 2015.

Chapter Eleven

Reflections

Edward M. Bruner

In reading this book, I'm surprised at how much I've learned from it about myself and about interpretive theory, constructivism, and the anthropology of tourism. I've come to see many aspects of my work in a new light. I'm exposed to nuances and new interpretations that I'd never thought of before as well as to new implications. Therefore, I want to express my appreciation to the editors of and contributors to this volume for what they have written and taught me. Not that I necessarily agree with it all. But the disagreements are points of entry for discussion. I've decided to comment on each paper in turn, responding selectively to points that may advance our understanding or that strike me as particularly interesting.

In their chapter, "Always in Process: Edward Bruner, American Anthropology, and the Study of Tourism," Nelson Graburn and Naomi Leite have written a brilliant historical analysis of the entire period in which I have conducted my research. They place my research in its larger societal context, beginning with my dissertation research and moving on to how I and anthropology subsequently changed. In other summaries of my work—for example, at my retirement party—the speakers began with my work starting in the 1980s even though I received my PhD in 1954. They began with the research that was more current and with which they were more familiar.

Most significantly, Graburn and Leite demonstrate that my constructivism began with my dissertation and continued throughout my career. There was a continuity in my theoretical analysis that even I did not fully realize. I thank the authors for pointing this out. They show that although anthropology generally was moving from formal positivism to constructivism, I had from the beginning, in my dissertation, been a constructivist, demonstrating that people do not just occupy roles, they construct them. Another contribution was that in the anthropology of tourism those involved were not just guests and hosts, or tourists and locals, but there was a third party, the reflexive anthropologist. In later years, I recognized the importance of performance in the creation of meaning, and with Victor Turner I saw that culture is always emergent and in process. I also turned to narration and agency. The main contribution with Turner, as developed in the *Anthropology of Experience*, was not to view culture as patterns or traits, but to view culture as experienced. In a study abroad program in 1983–1984, I learned from my co-faculty member on the trip, Barbara Kirshenblatt-Gimblett, more about performance theory and how to use it in practice.

Graburn and I followed a similar academic trajectory. We both went to the University of Chicago, I in the 1950s and he in the 1960s, worked with Fred Eggan, did dissertation research with Native Americans, and eventually turned to the anthropology of tourism. Graburn was familiar with my research from the beginning and was even influenced by some of my early papers. He is probably in the best position of anyone to evaluate my life work.

In "On the Emergence of Identity and Borderzones as Key Concepts," Kathleen Adams analyzes all my work from when I was a graduate student to the present. This complements the chapter by Graburn and Leite, which focuses on the early years, although Adams's is more abstract. She isolates the analytic issues of mobility, encounters, and shifting identities. Both, however, isolate my contributions and direct attention to the continuities in my approach. This astonishes me. For a restless nomadic anthropologist who has worked in different areas of the world—from Native America, to Indonesia, to Kenya, to Ghana, to China, to Israel, and in the United States—it's easy to lose sight of the continuities in my own research. As I switched to different areas of the world I found myself associating with new colleagues and new history and traditions. I once said somewhere that despite my peripatetic career, "I'm still the same Ed Bruner." Now, after reading these papers, I really believe it. There were some discrepancies in my different studies but I approached each research with basically the same theoretical framework, without being fully aware of it. It's gratifying to have other scholars point this out. Why didn't I know this myself? Possibly my dislike of polemics prevented me from attacking others, although a good hard dispute in the literature might have helped me to clarify and see the continuities in my own work. In any case, the papers in this volume do the job.

Julia Harrison's "'So in Effect I Was Studying Myself': Knowing (Our) Tourist Stories" is a chapter about stories—of how she entered anthropology, of her two years of backpacking, of international tourism, and of stories told to her by residents of the cottages in the lake area of Ontario. They are narratives told with sensibility and with attention to affect and the emotional undertones. As mentioned elsewhere in this volume, events may be approached from four angles, beginning with 1) imaginaries—that is what the storyteller conceives the event will be, the pretour understanding; 2) the actual trip as lived; 3) how the trip is experienced; and 4) the retellings after the event is over. Harrison focuses on retellings. This is relatively rare in the anthropological literature, and welcome.

Retellings may change with the audience and the context of the situation in which the story is told. Retellings are not an exact replica of the actual experience. But they do represent a synthesis of aspects of the actual event, and in most cases they provide more coherence and unity to the experience. They emphasize particular aspects of the event to the neglect of other aspects. They make sense of what might have been a more chaotic experience. I'd like to see an analysis of retellings over a period of years by the same persons to see how they change the telling. Or maybe the tourist has no one to tell about the trip. Possibly just after returning home there is a burst of tellings, and after that no one else is interested. The only alternative is to take another trip.

On to Walt Little's "'Whatever We Weave Is Authentic': Coproducing Authenticity in Guatemalan Tourism Textile Markets." Authenticity, I have always felt, is a folk concept, and not part of the theoretical conceptual apparatus of the anthropologist. When authenticity as a topic arises in the tourist situation, from whatever source, then the anthropologist is obligated to follow up and explore its meanings. These contexts can vary widely. Once in Bali the tourists and I saw a dance performance in the home of a member of the Javanese nobility. I pointed out how the dance and the whole setting was constructed for tourists and was not authentic. The tourists replied that they didn't care about authenticity, as it was a good show. In other cases, there were struggles over meaning that involved authenticity, or there were different narratives by different audiences, and at times these beliefs were deeply held. Situations vary.

I don't believe that tourists are necessarily traveling to distant places seeking to find their authentic selves, although some anthropologists I deeply respect believe this. But I was dealing with upscale, older, more wealthy tourists who knew who they were and were quite secure about their identity. They were just traveling for the fun and pleasure of it, and to educate themselves about the world.

Little's chapter is so wonderful because questions of authenticity arise so prominently, among both tourists and Guatemalan weavers, and these conversations are reported to us. We hear the voices of the participants: authenticity becomes alive, spoken by humans. Authenticity is indeed a constructed social coproduction, and Guatemalan weavers try to produce weavings that the tourists will accept as real Maya, weavings that the weaver might use herself. It is notable that the question of authenticity arises in situations in which a sale is to be made, and to question authenticity is sometimes used as a ploy to bring down the price. Little writes from the weavers' perspective rather than the tourists', on the basis of his twenty-five years of ethnographic experience among Mayas. The weavers' aim is to satisfy the buyers' conception of what is really Maya. Authenticity, his informant says, "is something gringos worry about."

Valerio Simoni's "Living in and Reaching beyond the Touristic Borderzone: A View from Cuba" expands the concept of the touristic borderzone. I agree with everything he says in this beautifully written chapter, but in contrasting my use of the touristic borderzone and the way he expands the concept, I feel an important distinction is lost. (I am sure Simoni understands this, but doesn't emphasize it.) I make the distinction between *performance time*, which takes place within the touristic borderzone, where locals perform for the tourists, and *real life*, after the performance is over and the tourists go back to their hotels and the locals go back to their own communities. Simoni argues that in his Cuban research, interaction in the borderzone may *become* real life. I agree entirely. The main difference in our use of the concept is the tourism context.

The tourism I studied was formal and structured. A tour agency sold the organized group tour, the printed itinerary was set in advance, and the tourists paid for the entire tour at the very beginning. In my experience, the tourists wanted to do everything on the itinerary, for fear that they may be missing something. If they were going by bus, which was often the case, after a set time the guide told everyone, "Back to the bus," and they went on to the next site. There was little opportunity for the locals and the tourists to develop any relations of intimacy. In the evening tourists did go back to their hotels where there were few locals except for performers and serving staff.

Simoni studied the informal sector of tourism, where local individuals established relations with tourists one on one. This changed the nature of the touristic borderzone. There was no set itinerary, and it was not a group tour. His work reminds me of that of an Illinois PhD student, Nicole Tami. Her thesis, "Romancing Strangers: The Intimate Politics of Beach Tourism in Kenya," was written in 2008. She worked with "beach boys" in Lamu, Mombasa, and Malinda. The beach boys—or, as she prefers, the beach workers—are free agents. They establish relationships with tourists by offering their service as guides, by going with them to the markets, and by offering boat rides—trying to create friendships.

These relations sometimes do lead to friendships and even sexual relations. The tourists in some cases take the beach workers back to their home country, perhaps Italy, Germany, Denmark, or Sweden. Some such relations may lead to marriage. So what happens in the borderzone is or becomes real life. Or the tourist may come to Lamu year after year to meet with the same beach worker. The beach workers themselves look forward to possible marriages or long-term relations. This situation has many similarities with Simoni's research data.[30] In each case, the researchers worked in the informal sector, even though there were structured formal group tours to these islands of the kind of tourism I have studied. Maybe there are other touristic situations in addition to these, and the borderzone concept could be further studied and clarified in each.[31]

Sally Ness's extraordinary chapter, "'I Can Feel Them Now, Even as I Write': Hiking Yosemite Falls with the Emergent Subjects of Tourism," has taught me more about my own theoretical orientation than I ever thought possible. She has relabeled my approach "pragmatic constructionism" and has shown the relationship between my key concepts of narrative, performance, interpretation, and meaning as emergent rather than predetermined. I agree with her sophisticated analysis of my work, which moves from hiking to philosophy with astonishing grace. Although anthropologists recognize the importance of the body in tourism, we may not incorporate it sufficiently into our research. Here the anthropologist's body *is* the tourist's body, and I'm amazed at Ness's stamina, courage, and strength. Toward the end of her hike I can feel her move from one step to another.

Kathleen Adams's chapter "'Being a Tourist in My (Own) Home': Negotiating Identity between Tourism and Migration in Indonesia" gives me an entirely new perspective on a place and people I knew well. She and I both studied inland mountain peoples in Indonesia: she the Toraja, and I the Toba Batak of Sumatra, both of whom are Christian minorities in Islamic Indonesia, have experienced extensive migration, and have many educated members. Thus via my long-term experience with Toba Batak communities, I have a grasp of their common experiences. I've had conversations with Toraja elders in Torajaland comparing their culture with that of the Batak. They were as interested in the subject as I was. My research with Toba Bataks started in 1957 in Sumatra, before smart phones and extensive use of the internet. So to read about Toraja now who have cell phones, take selfies, and have Facebook pages is dazzling to me. It changes the nature of fieldwork: not only does the anthropologist have these new tools, but the people we study also have the new media and use them. As they post on the internet, we can listen in or communicate with them electronically. This may all be familiar to younger anthropologists, but this old dog is astonished by it all.

Adams's chapter is also full of ideas and findings about Toraja return migration that I could apply or test out with the Batak, as probably one half of the Toba Batak population have left the ancestral homeland. I've neglected them in my own research. They maintain kinship ties, and many return home for weddings, funerals, dedications of clan monuments, and other events. I already know these people and many are my fictive kin. I'd love to study

Batak return migrants if I weren't ninety-three years old and unable to walk without a wheeled walker for support. But I can imagine the fun it would be. Adams has opened up for me a new world of research.

In a conversation with Hilly Geertz in Bali, Hilly made the point that we anthropologists make a great effort to distinguish ourselves from tourists, but from the Balinese perspective we are all tourists. In an endnote to his chapter, "Beyond Dialogue: Hospitality and the Transformation of Self in Southwestern Madagascar," David Picard says that the Malagasy do not distinguish between different types of foreign visitors. I think this is an important observation, in that it highlights the native perspective. Our view of guests and hosts, or of tourists and anthropologists, may distort how these actors are seen by the locals. We impose our Western categories.

But that is not Picard's main point. He pushes the boundaries of the touristic borderzone by showing how local actors employ strategies of hospitality based on their own master narratives. In my own work, I think I have neglected how local actors perceive and react to tourists and their belief systems. We may be imposing our own views too easily. And I am intrigued by how the spirits move in and out of these dialogical processes. This is a thoughtful and suggestive paper, based on really good ethnography.

Quetzil Castañeda's "'Ideologies at War' at Chichén Itzá" is an excellent ethnography of a contested site. It places Chichén Itzá in its wider context and also provides detailed examples of inside, local politics of the site. One could only write such a paper if one had an in-depth understanding of Chichén as a heterotopia, "an assemblage of meanings," a concept elaborated by Michel Foucault. Castañeda's writing is also deeply influenced by Bakhtin, an influence we share. For example, he writes that "each word is a polyphonic battlefield of ideologies." That is pure Bakhtin. As I read this chapter I realize that my own writing on contested sites has in some cases had too narrow a focus, perhaps because I focused particularly on the internal dynamics of the site itself.

With each chapter, the contributors to this book point in new directions and chart a path for the future of the anthropology of tourism. The seeds of tourism studies for the years and decades ahead are here in these pages.

NOTES

1. For further examples of this dynamic, see Brennan (2004), Dahles and Bras (1999), Meisch (1995), and Pruitt and LaFont (1995), among others.

2. Additional instances of blurring between the borderzone and "real life" have been documented in settings where local people and tourists are equally invested in finding commonality or creating lasting bonds. See, for example, Harrison (2003), Leite (2017), and Steel (2012).

Afterword

Beyond Anthropology

Ethnography in Tourism Studies

Mary Mostafanezhad and Margaret Byrne Swain

Lined with surfers, hammocks, and beachside bars, the coastline of San Juan del Sur, Nicaragua, has been dramatically reshaped by the growing surf tourism industry in one of the poorest countries in Latin America.[1] With a Nicaraguan Toña beer in one hand and a surfboard in the other, international tourists seek out sun, sea, and surf in the former sleepy fishing village. Interactions with tourists are now a daily part of life for most residents, who have seen sweeping shifts in the sociocultural and political-economic fabric of their lives.

The transitions in San Juan del Sur have been ethnographically investigated, but not by anthropologists. Rather, ethnographers with degrees and academic positions in tourism studies have used ethnographic methods—the sine qua non of anthropology—to describe and understand the meanings and implications of surf tourism development along this Nicaraguan coastline. Notably, their findings are not published only in academic articles. Their ethnographic descriptions are shared in popular media and on blogs and social media, complete with public participation in the form of likes, shares, and Tweets. This scenario reminds us that much has changed since Edward Bruner first cast his anthropological light on tourism three decades ago: ethnographic methods are now used by tourism-focused scholars from a range of disciplines, new communication technologies facilitate opportunities for sharing academic insights, and global tourism development is reshaping lives at unprecedented rates.

Writing from a transdisciplinary yet anthropologically rooted perspective, we observe that while Bruner's framings of tourism have indeed been influential, his most enduring influence is in the broadening of ethnography in tourism studies. This observation is significant not only for the ways in which anthropology has contributed to the birth of a more critical tourism studies, but for how tourism studies has reciprocated with its contribution to the development of ethnographic work that accounts for mobility in the ethnographic descriptions of culture. The rise of multi-sited ethnographies and ethnographic approaches that depart from classical ethnographic fieldwork helped pave the way for an ethnography unmoored from its anthropological dock.

Clearly articulated throughout this volume is Bruner's legacy in the anthropology of tourism. Yet tourism-focused scholars in disciplines as disparate as geography, business, and art often do not make explicit genealogical connections with his work. Perhaps unwittingly, scholars draw on many of his early conceptual developments, such as authenticity, encounter, performance, and narrative, without reference to Bruner. Thus, while Bruner's insights have helped fertilize the seeds of tourism studies, that tree now has many branches that extend well beyond its anthropological roots. This lack of theoretical synthesis between the anthropology of tourism and the transdisciplinary field of tourism studies reflects the durability of disciplinary walls that were cemented many decades ago, as well as the multifaceted nature of tourism as an industry and a practice.

In this afterword we reflect on Bruner's contributions beyond anthropology, as well as emerging trends in the field that often begin where Bruner left off. We consider how each of his core theoretical constructs—including the borderzone, embodiment, authenticity, and contested sites—have contributed to the building of new tourism scholarship. To conclude, we question the benefits and challenges of unhinging tourism ethnography from anthropology, and its emplacement in what is now described as critical tourism studies.

TOURISM ETHNOGRAPHY BEYOND ANTHROPOLOGY

Today tourism studies is a hodgepodge of transdisciplinary scholarship. It has been developed through a range of academic traditions, including anthropology but certainly not limited to it. In tourism studies departments, scholars debate the legitimacy of tourism studies as a distinct discipline and what is lost by surrendering to the diffuse nature of a transdisciplinary field (Echtner and Jamal 1997; Filep et al. 2015; Franklin and Crang 2001; Tribe 1997; 2010; 2016). Tourism studies scholars distinguish between tourism studies and tourism management, and young scholars seek to identify the core traditions of the field that developed from its more strictly defined disciplinary ancestors. Moreover, scholars further distinguish between critical and descriptive tourism studies.

Contemporary tourism scholars draw on ethnographic approaches to address questions of identity, meaning, and place. Many approach tourism from a qualitative and theoretical framework and identify with "critical tourism studies" as a distinct subfield. The development of this subfield is perhaps most visible in the two Critical Tourism Studies networks that are now global in scope and attract scholars from a wide range of disciplines. What began in Dubrovnik in 2005 as a primarily European-based tourism research network now includes two regional conferences: Critical Tourism Studies–North America and Critical Tourism Studies–Asia Pacific. Anthropologists meandering through one of the biennial conferences would perhaps be bewildered by the range of research methods that are described as ethnography. Although ethnography was invented by anthropologists, anthropologists seem to have lost their hold on the method that has long defined the discipline. Scholars from throughout the social sciences, humanities, and business schools now also describe their qualitative, fieldwork-based data collection methods as ethnography—some of which may be unrecognizable to the discerning anthropologist.

If we trace this transdisciplinary shift through Bruner's work, we are reminded that one of Bruner's early forays into the anthropology of tourism was a publication on Dennis O'Rourke's *Cannibal Tours* (E. Bruner 1989b). This is significant for how it focuses on ethnographic film, not by an anthropologist, but by an Australian cinematographer who draws on ethnographic modes of representation. The film is both a representation and a critique of anthropological inquiry into the so-called primitive world and Western fascinations with it. Through this and subsequent work, Bruner brought the ethnographer into view, not as the fly on the wall, but as part of the plot.

This history reminds us how one's subject position in the disciplines is not unlike one's position in the ethnographic encounter. Histories of disciplines are never fully written, and there are always alternative histories made invisible through acts of representation that naturalize inherently fictional borders (Marcus 2008). In the context of ethnographic truth, Clifford Geertz famously described how ethnographers would "never get to the bottom of things" through his story of an Englishman who sought to find out what was below the platform that the world rested on:

> There is an Indian story—at least I heard it as an Indian story—about an Englishman who, having been told that the world rested on a platform which rested on the back of an elephant which rested in turn on the back of a turtle, asked (perhaps he was an ethnographer; it is the way they behave), what did the turtle rest on? Another turtle. And that turtle? "Ah, Sahib, after that it is turtles all the way down." (Geertz 1973, 28–29)

Geertz's insights helped to usher in the "crisis of representation" in cultural anthropology, which triggered a new range of experimental ethnographic methods, as well as emergent themes in postmodern and postcolonial scholarship (Clifford 1986; Clifford and Marcus 1986; Marcus and Fischer 1986). These insights reshaped ethnographic practice and opened it up to new approaches that brought the researcher into the ethnography, and helped to encourage scholars from outside the disciplinary boundaries of anthropology to adopt ethnographic approaches to tourism research.

The offshoots of ethnography in other disciplines also meant that tourism ethnographies were not being written only by anthropologists. Geographers—the disciplinary cousins of anthropologists—began studying tourism during the golden age of flying in the 1950s and 1960s. Yet, while geographers were interested in questions of space, place, location, and environment, their work was often descriptive in scope, and they rarely applied ethnographic methods to their object of analysis (Timothy 2018). Geographers' themes, such as mobility, pilgrimage, and phenomenological studies of place, also crossed disciplinary boundaries in the late twentieth century, when the anthropologist's methodological tool kit was often found to provide the best mode of analysis through an ethnographic lens.

Since then, ethnography has been taught in a range of courses in the social sciences and beyond as a method to study the subtleties of tourism practices. Indeed, fieldwork, a rite of passage for students of anthropology, is now a widely appropriated tradition in numerous disciplines throughout the social sciences and humanities. Historians, sociologists, and tourism studies students regularly participate in a hybridized form of the ritualized event. As noted in the introduction to this volume, Bruner's ethnographic attention foregrounds questions of power, practice, and discourse. By interrogating the looking practices of anthropologists as well as tourists, and anthropologists *as* tourists, he contributed a more critical and reflexive ethnographic approach to tourism studies. Yet, as Ien Ang (2006) observes in the context of cultural studies—the theoretical driver of critical tourism studies—the "ethnographic turn" in her field barely resembles its anthropological ideal. In a similar vein, George Marcus notes that it is not just ethnography that has been unmoored from anthropology. He writes, "Even the concept of culture, while emblematic of what the discipline is interested in, is no longer viable analytically and has been appropriated everywhere and by everybody" (Marcus 2008, 1).

Unlike the peoples featured in the classical works of anthropologists such as E. E. Evans-Pritchard, Margaret Mead, and Bronisław Malinowski, Bruner's subjects of study did not stay put. Thus new methods became necessary for the anthropologist of tourism. In the microanalysis of the encounter, ethnography has been the primary method to account for the ways meaning is co-constituted through narrative, mobility, and imaginaries. The historical roots of the ethnographic study of tourism are well charted in anthropology, but they often omit the significance of influences from beyond anthropology's disciplinary borders, of which there are many.

It is not just the ethnographic study of tourism that has traveled beyond anthropology. Many anthropological constructs now form the theoretical roots of this transdisciplinary field. Drawing from multiple disciplines, tourism studies frequently synthesizes existing work and creates conceptual, theoretical, methodological, and translational innovations. It wasn't until 2005 that the "critical turn" in tourism studies was widely described as a distinct field of inquiry. Describing the critical tourism studies agenda, Nigel Morgan and colleagues (2018, 185; citing Ateljevic, Pritchard, and Morgan 2007) write that "we must dare tourism to develop conceptualizations that include multiple cultural differences and worldviews that reflect and recognize the plurality of human practices, positions, and insights." What Morgan and colleagues (2018) mean by "critical" here echoes many of the themes that had been central to anthropological, as well as geographical and sociological, studies of tourism since the late 1970s, thus reflecting the strength of the disciplinary borders that had kept tourism-focused scholars siloed in disciplinary enclaves. Other aspects of the critical turn include sustained reflection on tourism's relationship to the contemporary global economy, neoliberal development, social justice, and political power relations (Bianchi 2009), each of which had been addressed by tourism-focused geographers, anthropologists, and sociologists for at least several decades prior to the declaration of the critical turn in tourism studies. In what follows, we outline several classic analytical constructs in tourism studies that, while associated with Bruner's work in the anthropology of tourism, have also been shaped by transdisciplinary tourism scholarship.

BEYOND THE BORDERZONE

If the borderzone is a space in which so-called hosts and guests interact to co-create new cultural worlds (Bruner 1999, 461–62), then this space has become all the more complex. Increasing access to low-cost transportation and the growth of the new middle classes globally have contributed to increased diversity in the borderzone. No longer tethered to place (if they ever were), "hosts" around the world now have access to tourism experiences like never before. Indeed, as James Clifford noted in the late 1990s, to position Western cultures as "restless and expansive" while focusing on the "localization" of "rooted and immobile" natives is to overlook the long history of mobility among local cultures (Clifford 1997). Today, "hosts" travel both metaphorically through social media and physically on high-speed rails and via budget airlines. And while distinctions between leisure and work continue to be salient, they are also increasingly blurred by the rise of new forms of work that integrate with travel (such as digital nomadism). Today, we might suggest that the borderzone is more a rule than an exception.

Tourism-focused scholars have sought to make sense of the centrality of this movement of people through articulation with the development of mobilities studies. Mobilities studies takes the borderzone as its point of departure through its focus on the travel of people, ideas, and things. These ideas were perhaps foreshadowed by Clifford's earlier assertion, on the back cover of his book *Routes* (1997), that cultures are often "at home in motion." Thus within anthropology itself, Bruner's concepts were far from hegemonic and were sometimes overlooked by scholars who sought to describe similar phenomena. For instance, Clifford (1997) was similarly interested in issues of representation, power, and travel, and he developed remarkably similar concepts, such as the contact zone, to describe the space of encounter between residents and mobile people in postcolonial spaces.

Drawing on many of the concepts and themes from tourism studies, mobilities has developed its own disciplinary positioning, replete with its own following, conferences, and journals. The first issue of *Mobilities* journal was published in 2006. Today, anthropologists, geographers, sociologists, and others draw on mobilities scholarship in order to describe mobile populations as well as their ideas and things (Hannam, Sheller, and Urry 2006; Sheller and Urry 2006). New theoretical constructs have developed from this framing and contributed novel insights into the study of mobile populations.

While this work has expanded outward, it has also looked inward to consider the role of the subject position of the researcher. As a result, autoethnography is now widely used by tourism scholars to address subjectivity in tourism research. Critical tourism studies scholars often identify the significance of a reflexive understanding of political positioning. They question the meaning of tourism research and what it means to be an academic activist. Today, scholarship that takes the personal as political is widespread in critical tourism studies. Thus the borderzone has become deeply personal and political; for many, it is a significant site from which to engage with questions of equality and social justice.

BEYOND EMBODIMENT

The embodied nature of tourism reflects the ways that narratives are internalized, driving desire for particular kinds of experiences. Ever since Louis

Turner and John Ash's *The Golden Hordes* (1975), tourism scholars have addressed tourism as an embodied experience replete with emotional and affective attachments. This work was echoed in mainstream anthropology, which often sought to address the interplay between social structure, sentiment, and ideology (Abu-Lughod 2000). Today tourism scholars have built on these insights to integrate affect and intersectionality. Issues of race and ethnicity, inequality, gender, and environmental concerns now take center stage in the theater of tourism encounters. For instance, in their 2005 Critical Tourism Studies keynote address, Margaret Swain and Derek Hall (2007) examined gender and embodiment as critical to the analysis of tourism experience. Indicative of the transdisciplinary nature of their work, Swain and Hall also drew from feminist works on positionality and the subjectivity of the researcher. Feminist ethnographic approaches have been significant for the ways they have shaped how we understand self/other dichotomies and have promoted reflexive and first-person accounts. They have also contributed to our awareness of power in ethnographic encounters and have been at the forefront of shifting ethnographic norms in anthropology and beyond. Indeed, feminist ethnographers were the first to bring women's life stories into ethnographic view, while also creating interdisciplinary space for ethnographic approaches (Enslin 1994).

BEYOND AUTHENTICITY

The significance of ethnography in the authenticity debates can perhaps not be overstated. As Bruner has long argued, if we accept culture as an emergent, unbounded process, then our descriptions of authenticity shift from the realization of authenticity to an acceptance of the always emergent authenticity in experience. Arguably the most widely addressed issue in tourism studies, authenticity has been at the center of tourism scholarship. Bruner was among the first to argue that an analytical distinction between "authentic" and "inauthentic" tourism experiences is futile. Yet these insights have continued to transform. Conceptualizations of authenticity have gone through a range of iterations over the years, marked by terms such as staged (MacCannell 1973; Chhabra, Healy, and Sills 2003), performative (Knudsen and Waade 2010a), embodied (Buchmann, Moore, and Fisher 2010), experiential and emergent (Cohen 1988), negotiated (Cohen 1988; K. Adams 1997), and existential (Wang 1999). There have also been explicit efforts as well to go beyond (S. Cole 2007; Conran 2006), to rethink (Wang 1999; Zhu 2012), to reconceptualize (Reisinger and Steiner 2005), and to reinvest (Knudsen and Waade 2010b) authenticity as a meaningful concept in tourism research.

Today, authenticity of voice is also a core concern of scholars in both anthropology (especially public anthropology) and critical tourism studies. A growing interest in collaborative research is reflected in a range of ethnographic approaches that no longer speak of informants, but of collaborators or "situated co-researchers" (Swain 2016). Collaborative ethnography is "an approach to ethnography that deliberately and explicitly emphasizes collaboration at every point in the ethnographic process, without veiling it—from project conceptualization, to fieldwork, and, especially, through the writing process" (Lassiter 2005, 16). This approach has been highly influential in some niches of tourism studies, especially among scholars focused on community-based and indigenous tourism enterprises (Hales et al. 2018).

Time, long a theoretical interest of anthropologists, has also been brought to the foreground of tourism studies of authenticity. MacCannell (1973) was one of the first to demonstrate the time travel aspect of tourism in the context of so-called primitive cultures, which were perceived to be more authentic for their "premodern" existence. This work has been reconsidered in a range of different contexts. Scholars have scaled up these analyses to examine the chronopolitics, or politics of time, of tourism (Norum and Mostafanezhad 2016). This work describes the ways in which representations of temporality in tourism draw upon and perpetuate existing stereotypes, as well as mediate geopolitical discourse and practice. The actual temporality of the tourist experience is also significant here. Modes of self-actualization through tourism provide important insights into perceptions of the authenticity of experience. These insights are relevant not just for tourists, but also for ethnographers. The time ethnographers spend in the field has long been the yardstick by which the authenticity of their experience is—at least in part—measured.

BEYOND CONTESTED SITES

An ethnographic focus on contested sites has challenged scholars to address multiple interpretations of history and place. Contemporary tourism scholars address the inevitability of contestation over the representation of place. The perhaps unavoidable contestation over sites has left the door open for a range of ethnographic approaches to place-making practices. Tourism scholars also address the spatial practices through which tourist sights and sites take shape. This contestation is perhaps most visibly addressed through the emergent theme of tourism geopolitics (Dowler 2013; Mostafanezhad and Norum 2016), in scholarship that scales up analyses of contested sites from place-based studies to multiscalar accounts of how tourism practice both reflects and mediates geopolitical practice. This work has also influenced an environmental geopolitics of tourism that addresses how narratives of the environment support or contest geopolitical power and place, as well as the spatial aspects of risk and security as they relate to the environment (O'Lear 2018, 2).

The coproduction of contested sites through storytelling has been a topic of significant interest to tourism studies scholars. For instance, Keith Hollinshead and Rukeya Suleman (2017, 200) explore transdisciplinary applications of this concept to the ordinary activities of tourism as well as the "mundane/banal orientations" of tourism studies. They describe acts of worldmaking, or the imaginative processes and activities that privilege dominant representations of populations, places, and pasts over others. This framing of the relationship between representation, place, and imagination is also captured by the concept of geographical imaginations. Geographical imaginations refers to our mental images or perspectives of peoples and places that are inherently imbued within the politics of representation (Agnew 1989; Gregory 1994; Harvey 1990; Schwartz and Ryan 2003). Linked to colonial and postcolonial contexts, geographical imaginations are enlisted by tourists and the tourism industry to distinguish between self and other as well as between places.

These performances of differences in tourism are significant in that, as Edward Said observed, "Just as none of us is outside or beyond geography, none of us is completely free from the struggle over geography" (1993, 7; cited in McFarlane 2004, 894). Colin McFarlane describes how "that struggle is complex and interesting because it is not only about soldiers and cannons but also about ideas, about forms, about images and imaginings" (2004, 894). Further enlarging this field is the analytic of tourism imaginaries, or the "powerful deconstruction device of ideological, political, and sociocultural stereotypes and clichés" of tourism (Salazar 2010, 7; 2012, 865; see also Salazar and Graburn 2014). Thus geographers' geographical imaginations, anthropologists' tourism imaginaries, and tourism studies scholars' worldmaking practices have collectively sought to deepen our understanding of the drivers and consequences of contested sites in tourism.

CONCLUSION

All too often, tourism scholars struggle to have the object of their research taken seriously. They strive diligently to describe the significance of their work beyond the tourism encounter itself. Tourism-focused anthropologists can no longer afford to ignore the theoretical insights of scholars from outside the discipline, and scholars outside the discipline cannot claim to invent concepts and theoretical persuasions that have long existed in anthropology.

Rather, as in other transdisciplinary subfields, cross-pollination is now the rule rather than the exception. The institutional drivers of academia have long maintained separate fields of scholarship. As an increasing number of anthropologists are employed in academic departments other than anthropology, the turn toward transdisciplinary tourism studies is not only theoretically driven by but also reflects the reorganization of academic institutions. Thus, many tourism scholars might agree with George Marcus, who argues that the reorganization of cultural anthropology "has left the center of the discipline intellectually weak relative to the vitality of its diverse interdisciplinary and even nonacademic engagements." (Marcus 2008, 1)

Rather than policing the boundaries of ethnography, we champion theoretical innovation that blurs the disciplinary boundaries in tourism studies. Tourism ethnographers, we argue, have not only borrowed from anthropology's methodological tool kit, but also contributed new instruments to help scholars study mobile cultural forms. Like all good things, these benefits do not come without a cost. The unmooring of ethnography from socio-cultural anthropology also presents numerous challenges. The emplacement of ethnography in what is now a decidedly transdisciplinary tourism studies has meant that some of the refinements of ethnographic approaches and ethical contours that anthropologists have long labored over are lost in translation. In many ways, transdisciplinary fields such as tourism studies have also developed their own narrow perspectives that obscure the historical disciplinary fragments from which their field has been cobbled together. Yet as the permeability of disciplinary borders continues to increase, it behooves us to take note of the growth of new branches as well as the roots from which they arise.

NOTE

1. Cinnamon Janzer, "Surf Tourism is a Fun Vacation That's Killing Central American Economies." *Pacific Standard,* December 25, 2017, psmag.com/economics/surf-tourism-is-killing-central-american-economies (accessed October 25, 2018).

References Cited

Abram, Simone, Jacqueline Waldren, and Don MacLeod, eds. 1997. *Tourists and Tourism: Identifying with People and Places.* New York: Berg.

Abu-Lughod, Lila. 2000. "Locating Ethnography." *Ethnography* 1 (2): 261–67.

Acciaioli, Greg. 1985. "Culture as Art: From Practice to Spectacle in Indonesia." *Canberra Anthropology* 8 (1–2): 148–72.

Adams, Kathleen M. 1984. "Come to Tana Toraja, 'Land of the Heavenly Kings': Travel Agents as Brokers in Ethnicity." *Annals of Tourism Research* 11 (3): 469–85.

Adams, Kathleen M. 1991. "Distant Encounters: Travel Literature and the Shifting Image of the Toraja of South Sulawesi, Indonesia." *Terrae Incognitae* 23: 1–12.

Adams, Kathleen M. 1997. "Ethnic Tourism and the Renegotiation of Tradition in Tana Toraja (Sulawesi, Indonesia)." *Ethnology* 36 (4): 309–20.

Adams, Kathleen M. 1998. "Domestic Tourism and Nation-Building in South Sulawesi." *Indonesia and the Malay World* 26 (75): 77–96.

Adams, Kathleen M. 2006. *Art as Politics: Re-crafting Identities, Tourism, and Power in Tana Toraja, Indonesia.* Honolulu: University of Hawai'i Press.

Adams, Kathleen M. 2012. "Ethnographic Methods." In *Handbook of Research Methods in Tourism: Qualitative and Quantitative Methods*, edited by Larry Dwyer, Alison Gill, and Neelu Seertaram, 339–51. Northampton, MA: Edward Elgar.

Adams, Kathleen M. 2015a. "Families, Funerals, and Facebook: Re-imag(in)ing and Curating Toraja Kin in Translocal Times." *TRaNS: Trans-Regional and -National Studies of Southeast Asia* 3 (2): 239–66.

Adams, Kathleen M. 2015b. "Identity, Heritage, and Memorialization: The Toraja *Tongkonan* of Indonesia." In *Writing Material Culture History*, edited by Anne Gerritsen and Giorgio Riello, 93–100. London: Bloomsbury.

Adams, Kathleen M. 2016. "Tourism and Ethnicity in Insular Southeast Asia: Eating, Praying, Loving, and Beyond." *Asian Journal of Tourism Research*, 1 (1): 1–28.

Adams, Richard N. 1970. "Brokers and Career Mobility Systems in the Structure of Complex Societies." *Southwestern Journal of Anthropology* 26 (4): 315–27.

Adams, Vincanne. 1992. "Tourism and Sherpas, Nepal: Reconstruction of Reciprocity." *Annals of Tourism Research* 19 (3): 534–54.

Adams, Vincanne. 1996. *Tigers of the Snow and Other Virtual Sherpas: An Ethnography of Himalayan Encounters.* Princeton, NJ: Princeton University Press.

Aerni, Mary Jean. 1972. Letter to the editor. *Current Anthropology* 13 (2): 162.

Agnew, John A. 1989. *The Power of Place: Bringing Together Geographical and Sociological Imaginations*. Sydney: Allen & Unwin Australia.

Albright, Evan J. 2015. *The Man Who Owned a Wonder of the World: The Gringo History of Mexico's Chichén Itzá*. Bourne, MA: Bohlin Carr.

Allen, Jafari S. 2007. "Means of Desire's Production: Male Sex Labor in Cuba." *Identities: Global Studies in Culture and Power* 14 (1–2): 183–202.

Aloula, Malek. 1986. *The Colonial Harem.* Minneapolis: University of Minnesota Press.

Amit, Vered, ed. 1999. *Constructing the Field: Ethnographic Fieldwork in the Contemporary World.* London: Routledge.

Andrews, Hazel, Takamitsu Jimura, and Laura Dixon, eds. 2019. *Tourism Ethnographies: Ethics, Methods, Application, and Reflexivity.* London: Routledge.

Ang, Ien. 2006. *Living Room Wars: Rethinking Media Audiences*. New York: Routledge.

Annis, Sheldon. 1987. *God and Production in a Guatemalan Town*. Austin: University of Texas.

Ansari, Ghaus. 1971. Letter to the editor. *Current Anthropology* 11 (1): 1.

Anzaldúa, Gloria E. 1987. *Borderlands/La Frontera: The New Mestiza*. San Francisco: Spinsters/Aunt Lute Foundation.

Appadurai, Arjun, ed. 1986. *The Social Life of Things: Commodities in Cultural Perspective*. Cambridge, UK: Cambridge University Press.

Arensberg, Conrad. 1937. *The Irish Countryman: An Anthropological Study.* Gloucester, MA: Peter Smith.

Armstrong-Fumero, Fernando. 2013. *Elusive Unity: Factionalism and the Limits of Identity Politics in Yucatán, Mexico*. Boulder: University of Colorado Press.

Astuti, Rita. 1995. "'The Vezo Are Not a Kind of People': Identity, Difference, and 'Ethnicity' among a Fishing People of Western Madagascar." *American Ethnologist* 22 (3): 464–82.

Ateljevic, Irena, Annette Pritchard, and Nigel Morgan, eds. 2007. *The Critical Turn in Tourism Studies*. New York: Routledge.

Azarya, Victor. 2004. "Globalization and International Tourism in Developing Countries: Marginality as a Commercial Commodity." *Current Sociology* 52: 949–67.

Babb, Florence E. 2010. *The Tourism Encounter: Fashioning Latin American Nations and Histories.* Palo Alto, CA: Stanford University Press.

Babcock, Barbara. 1986. *The Pueblo Storyteller: Development of a Figurative Ceramic Tradition*. Tucson: University of Arizona Press.

Bakhtin, Mikhail. 1981. *The Dialogic Imagination: Four Essays by M. M. Bakhtin*, edited by Michael Holquist; translated by Caryl Emerson and Michael Holquist. Austin: University of Texas Press.

Barba, Eugenio. 1993. *The Paper Canoe: A Guide to Theatre Anthropology*. London: Routledge.

Barthes, Roland. 1974. *S/Z*, translated by Richard Miller. New York: Hill and Wang.

Bartlett, Frederic. 1932. *Remembering: A Study in Experimental and Social Psychology*. Cambridge, UK: Cambridge University Press.

Bateson, Gregory. 1972. *Steps to an Ecology of Mind.* New York: Basic Books.

Beer, Jenny. 1993. "Packaged Experience: Japanese Tours to Southeast Asia." PhD diss., University of California, Berkeley.

Behar, Ruth, and Deborah Gordon, eds. 1995. *Women Writing Culture.* Berkeley: University of California Press.

Benveniste, Émilie. 1969. *Le vocabulaire des institutions indoeuropeennes: Sommaires, tableau et index établis*. Paris: Editions de minuit.

Berger, Peter, and Thomas Luckmann. 1967. *The Social Construction of Reality*. Garden City, NY: Doubleday.

Berlant, Lauren. 2011. *Cruel Optimism*. Durham, NC: Duke University Press.

Bhabha, Homi. 1994. *The Location of Culture*. London: Routledge.

Bianchi, Raoul V. 2009. "The 'Critical Turn' in Tourism Studies: A Radical Critique." *Tourism Geographies* 11 (4): 484–504.

Bloch, Maurice. 2011. "The Blob." *Anthropology of This Century* [online journal]. Issue 1. http://aotcpress.com/articles/blob/

Blumer, Herbert. 1969. *Symbolic Interactionism.* Englewood Cliffs, NJ: Prentice-Hall.

Boas, Franz. 1896. "The Limitations of the Comparative Method in Anthropology." *Science* 4 (103): 901–8.

Boon, James A. 1982. *Other Tribes, Other Scribes.* Cambridge, UK: Cambridge University Press.

Boon, James A. 1999. *Verging on Extra-Vagance*. Princeton, NJ: Princeton University Press.

Boorstin, Daniel. 1962. *The Image: A Guide to Pseudo-Events in America.* New York: Harper Colophon.

Bowman, Glenn. 1996. "Passion, Power, and Politics in a Palestinian Tourist Market." In *The Tourist Image: Myths and Myth Making in Tourism*, edited by Tom Selwyn, 84–103. Chichester, UK: John Wiley.

Breglia, Lisa C. 2006. *Monumental Ambivalence.* Austin: University of Texas Press.

Breglia, Lisa C. 2016. "The Business of Wonder: Public Meets Private at the World Heritage Site of Chichén Itzá." In *World Heritage on the Ground*, edited by Christoph Brumann and David Berliner, 193–216. New York: Berghahn.

Brennan, Denise. 2004. *What's Love Got to Do with It? Transnational Desires and Sex Tourism in the Dominican Republic*. Durham, NC: Duke University Press.

Brulotte, Ronda. 2012. *Between Art and Artifact: Archaeological Replicas and Cultural Production in Oaxaca, Mexico*. Austin: University of Texas Press.

Bruner, Edward M. 1953. "Assimilation among Fort Berthold Indians." *American Indian* 6 (4): 21–29.

Bruner, Edward M. 1954. "A Study of Cultural Change and Persistence in a Mandan-Hidatsa Indian Village." PhD diss., University of Chicago.

Bruner, Edward M. 1955. "Two Processes of Change in Mandan-Hidatsa Kinship Terminology." *American Anthropologist* 57 (4): 840–50. Reprinted in Nelson Graburn, ed., *Readings in Kinship and Social Structure* (New York: Harper & Row, 1971), 119–26.

Bruner, Edward M. 1956a. "Cultural Transmission and Cultural Change." *Southwestern Journal of Anthropology* 12 (2): 191–99.

Bruner, Edward M. 1956b. "Primary Group Experience and the Process of Acculturation." *American Anthropologist* 59 (4): 605–23.

Bruner, Edward M. 1961. "Urbanization and Ethnic Identity in North Sumatra." *American Anthropologist* 63 (3): 508–21.

Bruner, Edward M. 1963. "Review of *Encounters: Two Studies in the Sociology of Interaction* by Erving Goffman." *American Anthropologist* 65: 1416–17.

Bruner, Edward M. 1964. "The Psychological Approach in Anthropology." In *Horizons of Anthropology*, edited by Sol Tax, 71–80. Chicago: Aldine.

Bruner, Edward M. 1973. "The Missing Tins of Chicken: A Symbolic Interactionist Approach to Culture Change." *Ethos* 1 (2): 219–38.

Bruner, Edward M. 1984a. "Introduction: The Opening Up of Anthropology." In *Text, Play, and Story: The Construction and Reconstruction of Self and Society*, edited by Edward M. Bruner, 1–16. Washington, DC: American Ethnological Society.

Bruner, Edward M., ed. 1984b. *Text, Play, and Story: The Construction and Reconstruction of Self and Society.* Washington, DC: American Anthropological Association.

Bruner, Edward M. 1986a. "Anthropology and Human Studies." *Cultural Anthropology* 1 (1): 121–24.

Bruner, Edward M. 1986b. "Ethnography as Narrative." In *The Anthropology of Experience*, edited by Victor Turner and Edward M. Bruner, 139–55. Urbana: University of Illinois Press.

Bruner, Edward M. 1986c. "Introduction: Experience and Its Expressions." In *The Anthropology of Experience*, edited by Victor Turner and Edward M. Bruner, 3–30. Urbana: University of Illinois Press.

Bruner, Edward M. 1989a. "About the Author." *Cultural Anthropology* journal website, https://culanth.org/articles/371-of-cannibals-tourists-and-ethnographers, accessed August 3, 2017.

Bruner, Edward M. 1989b. "Of Cannibals, Tourists, and Ethnographers." *Cultural Anthropology* 4 (4): 438–45.

Bruner, Edward M. 1989c. "Tourism, Creativity, and Authenticity." *Studies in Symbolic Interaction* 10: 109–14.

Bruner, Edward M. 1990. "The Scientists vs. the Humanists." *Anthropology News* 31 (2): 28.

Bruner, Edward M. 1991. "Transformation of Self in Tourism." *Annals of Tourism Research* 18 (2): 238–50.

Bruner, Edward M. 1993a. "Introduction: The Ethnographic Self and the Personal Self." In *Anthropology and Literature*, edited by Paul Benson, 1–26. Urbana: University of Illinois Press.

Bruner, Edward M. 1993b. "Lincoln's New Salem as a Contested Site." *Museum Anthropology* 17 (3): 14–25. Revised and reprinted as chapter 4 in Bruner 2005a.

Bruner, Edward M. 1994. "Abraham Lincoln as Authentic Reproduction." *American Anthropologist* 96 (2): 397–415. Revised and reprinted as chapter 5 in Bruner 2005a.

Bruner, Edward M. 1995. "The Ethnographer/Tourist in Indonesia." In *International Tourism: Identity and Change*, edited by Marie-Françoise Lanfant, J. B. Allcock, and Edward M. Bruner, 224–41. London: Sage.

Bruner, Edward M. 1996a. "Tourism in Ghana: The Representation of Slavery and the Return of the Black Diaspora." *American Anthropologist* 98 (2): 290–304. Revised and reprinted as chapter 3 in Bruner 2005a.

Bruner, Edward M. 1996b. "Tourism in the Balinese Borderzone." In *Displacement, Diaspora, and Geographies of Identity*, edited by Smadar Lavie and Ted Swedenburg, 157–70. Durham, NC: Duke University Press. Revised and reprinted as chapter 7 in Bruner 2005a.

Bruner, Edward M. 1999. "Return to Sumatra: 1957, 1997." *American Ethnologist* 26 (2): 461–77. Revised and reprinted as chapter 9 in Bruner 2005a.

Bruner, Edward M. 2001. "The Maasai and the Lion King: Authenticity, Nationalism, and Globalization in African Tourism." *American Ethnologist* 28 (4): 881–908. Revised and reprinted as chapter 2 in Bruner 2005a.

Bruner, Edward M. 2005a. *Culture on Tour: Ethnographies of Travel*. Chicago: University of Chicago Press.

Bruner, Edward M. 2005b. "Introduction: Travel Stories Told and Retold." In *Culture on Tour: Ethnographies of Travel*, by Edward M. Bruner, 1–29. Chicago: University of Chicago Press.

Bruner, Edward M. 2005c. "The Role of Narrative in Tourism." Paper presented at "On Voyage: New Directions in Tourism Theory," Berkeley, CA, October 7–8.

Bruner, Edward M. 2005d. "Taman Mini: Self-Constructions in an Ethnic Theme Park in Indonesia." In *Culture on Tour: Ethnographies of Travel*, by Edward M. Bruner, 211–30. Chicago: University of Chicago Press.

Bruner, Edward M. 2005e. "Through the Looking Glass: Reflections on an Anthropological Life." *Anthropology and Humanism* 30 (2): 201–7.

Bruner, Edward M. 2005f. "Tourism Fieldwork." *Anthropology News* 46 (5): 16, 19.

Bruner, Edward M. 2010. "Remembering My Jewish Father." *Anthropologica* 52 (1): 191–95.

Bruner, Edward M. 2011. "Afterword: The Tour as Imagined, Lived, Experienced, and Told." In *Great Expectations: Imagination, Anticipation, and Enchantment in Tourism*, edited by Jonathan Skinner and Dimitrios Theodossopoulos, 196–201. Oxford: Bergahn Books.

Bruner, Edward M. 2012. "Around the World in Sixty Years: From Native America to Indonesia to Tourism and Beyond." In *The Restless Anthropologist: New Fieldsites, New Visions*, edited by Alma Gottlieb, 138–58. Chicago: University of Chicago Press.

Bruner, Edward M. 2014. "The Ageing Anthropologist." *Anthropology and Humanism* 39 (1): 27–31.

Bruner, Edward M., and Judith O. Becker, eds. 1979. *Art, Ritual, and Society in Indonesia*. (Southeast Asia Series, 53.) Athens, OH: Ohio University Center for International Studies.

Bruner, Edward M., and Phyllis Gorfain. 1984. "Dialogic Narration and the Paradoxes of Masada." In *Text, Play, and Story*, edited by Edward M. Bruner, 56–79. Long Grove, IL: Waveland. Revised and reprinted as chapter 6 in Bruner 2005a.

Bruner, Edward M., and Jane Paige Kelso. 1980. "Gender Differences in Graffiti: A Semiotic Perspective." *Women's Studies International Quarterly* 3: 239–52.

Bruner, Edward M., and Barbara Kirshenblatt-Gimblett. 1987. "East Africa: Tourist Performances and Representations." *Problemy Turystyki (Problems of Tourism)* 10 (3): 25–29.

Bruner, Edward M., and Barbara Kirshenblatt-Gimblett. 1989. "Tourism." In *International Encyclopedia of Communications*, vol. 4, edited by Eric Barnouw, 249–53. New York: Oxford University Press.

Bruner, Edward M., and Barbara Kirshenblatt-Gimblett. 1994. "Maasai on the Lawn: Tourist Realism in East Africa." *Cultural Anthropology* 9 (4): 435–70. Revised and reprinted as chapter 1 in Bruner 2005a.

Bruner, Jerome S. 1994. "The 'Remembered' Self." In *The Remembering Self: Construction and Accuracy in the Self-Narrative*, edited by Ulric Neisser and Robyn Fivush, 41–54. Cambridge, UK: Cambridge University Press.

Bruner, Jerome S. 2003. *Making Stories: Law, Literature, Life.* Cambridge, MA: Harvard University Press.

Buchmann, Anne, Kevin Moore, and David Fisher. 2010. "Experiencing Film Tourism: Authenticity and Fellowship." *Annals of Tourism Research* 37 (1): 229–48.

Bunten, Alexis C. 2008. "Sharing Culture or Selling Out? Developing the Commodified Persona in the Heritage Industry." *American Ethnologist* 35 (3): 380–95.

Bunten, Alexis C. 2015. *So, How Long Have You Been Native? Life as an Alaska Native Tour Guide*. Lincoln, NE: University of Nebraska Press.

Bunten, Alexis C., and Nelson Graburn, eds. 2018. *Indigenous Tourism Movements.* Toronto: University of Toronto Press.

Cabezas, Amalia L. 2006. "The Eroticization of Labor in Cuba's All-Inclusive Resorts: Performing Race, Class, and Gender in the New Tourist Economy." *Social Identities* 12 (5): 507–21.

Campbell, Sarah. 2018. "Identity in Performance in Contemporary Yucatec Maya Theatre." PhD diss., Indiana University.

Candea, Matei, and Giovanni Da Col. 2012. "The Return to Hospitality." *Journal of the Royal Anthropological Institute* 18: 1–19.

Castañeda, Quetzil. 1996. *In the Museum of Maya Culture: Touring Chichén Itzá.* Minneapolis: University of Minnesota Press.

Castañeda, Quetzil E. 1998. "On the Correct Training of Indios at the Handicraft Market at Chichén Itzá." *Journal of Latin American Anthropology* 2 (2): 106–43.

Castañeda, Quetzil E. 2001. "Approaching Ruins: A Photo-Ethnographic Essay on the Busy Intersections of Chichén Itzá." *Visual Anthropology Review* 16 (2): 43–70.

Castañeda, Quetzil. 2004a. "Art-Writing in the Maya Art World of Chichén Itzá: Transcultural Ethnography and Experimental Fieldwork." *American Ethnologist* 31 (1): 21–42.

Castañeda, Quetzil E. 2004b. "'We Are Not Indigenous!': An Introduction to the Maya Identity of Yucatán." *Journal of Latin American Anthropology* 9 (1): 36–63.

Castañeda, Quetzil E. 2005a. "Between Pure and Applied Research: Experimental Ethnography in a Transcultural Tourist Art World." *National Association for the Practice of Anthropology Bulletin* 23 (1): 87–118.

Castañeda, Quetzil E. 2005b. "The Carnegie Mission and Vision of Science: Institutional Contexts of Maya Archaeology and Espionage." *Histories of Anthropology Annual* 1: 27–60.

Castañeda, Quetzil E. 2008. "'The Ethnographic Turn': Research Positioning and Reflexivity in Ethnographic Archaeologies." In *Ethnographic Archaeologies,* edited by Quetzil Castañeda and Chris Matthews, 25–62. Lanham, MD: AltaMira.

Castañeda, Quetzil E. 2009a. "Aesthetics and Ambivalence of Maya Modernity." In *Crafting Maya Identity*, edited by Jeff Kowalski, 133–52. DeKalb, IL: Northern Illinois University Press.

Castañeda, Quetzil E. 2009b. "Heritage and Indigeneity: Transformations in the Politics of Tourism." In *Cultural Tourism in Latin America: The Politics of Space and Imagery*, edited by M. Baud and A. Ypeji, 263–96. Leiden: Brill.

Castañeda, Quetzil E. 2009c. "Notes on the Work of Heritage in the Age of Archaeological Reproduction." In *Archaeologies and Ethnographies: Iterations of the Past*, edited by Lena Mortensen and Julie Hollowell, 109–19. Gainesville: University Press of Florida.

Castañeda, Quetzil E. 2010. "'Conjunctivitis': Notes on Historical Ethnography, Paradigms, and Social Contexts of Archaeology." In *Prophet, Pariah, and Pioneer: Walter W. Taylor and Dissension in American Archaeology*, edited by Allan Maca, Jonathan Reyman, and William Folan, 333–56. Boulder: University Press of Colorado.

Castañeda, Quetzil E. 2017. "In Search of Adventure: Simmel and Similar Travels among (Maya) Ruins." *Anthropological Quarterly* 90 (4): 931–66.

Castañeda, Quetzil E., and Jennifer Mathews. 2013. "Archaeology Meccas of Tourism: Exploration, Protection, Exploitation." In *Tourism and Archaeology: Sustainable Meeting Grounds*, edited by Cameron Walker and Neil Carr, 37–64. Walnut Creek, CA: Left Coast.

Castillo Cocom, Juan A. 2005. "'It Was Simply Their Word': Yucatec Maya PRInces in YucaPAN and the Politics of Respect." *Critique of Anthropology* 25 (2): 131–55.

Castillo Cocom, Juan A. 2007. "Maya Scenarios: Indian Stories in and out of Context." *Kroeber Anthropological Society Papers* 96: 13–35.

Castillo Cocom, Juan A. with Saúl Ríos Luviano. 2012. "Hot and Cold Politics of Indigenous Identity: Legal Indians, Cannibals, Words, More Words, More Food." *Anthropological Quarterly* 85 (1): 229–56.

Castillo Cocom, Juan A., Timoteo Rodriguez, and McCale Ashenbrener. 2017. "Ethnoexodus: Escaping Mayaland." In *"The Only True People": Linking Mayan Identities Past and Present*, edited by B. J. Beyette and L. LeCount, 42–72. Boulder: University of Colorado Press.

Causey, Andrew. 1999. "Stealing a Good Idea: Innovation and Competition among Toba Batak Woodcarvers." *Museum Anthropology* 23 (1): 33–46.

Causey, Andrew. 2003. *Hard Bargaining in Sumatra: Western Travelers and Toba Bataks in the Marketplace of Souvenirs*. Honolulu: University of Hawai'i Press.

Certeau, Michel de. 1984. *The Practice of Everyday Life*, translated by Steven F. Rendall. Berkeley: University of California Press.

Chambers, Erve. 2009. *Native Tours: The Anthropology of Travel and Tourism,* 2nd ed. Prospect Heights, IL: Waveland.

Chandler, Jerry. 2012. "RE: A Question of Artistic Semio-genesis." Comments posted on the Peirce-L listserv, accessed October 15, 2012.

Chapman, Charlotte Gower. 1971 [1935]. *Milocca: A Sicilian Village*. Cambridge, MA: Schenkman Publishing.

Chaudat, Pierre. 1993. "La possession vorombe dans la région de Tuléar." *Galaxie Anthropologie* 2–3: 130–36.

Chhabra, Deepak, Robert Healy, and Erin Sills. 2003. "Staged Authenticity and Heritage Tourism." *Annals of Tourism Research* 30 (3): 702–19.

Chronis, Athinodoros. 2005. "Co-Constructing Heritage at the Gettysburg Storyscape." *Annals of Tourism Research* 32: 386–406.

Chronis, Athinodoros. 2012. "Between Place and Story: Gettysburg as Tourism Imaginary." *Annals of Tourism Research* 39: 1797–816.

Clarke, Richard. 2000. "Self-Presentation in a Contested City: Palestinian and Israeli Political Tourism in Hebron." *Anthropology Today* 16 (5): 12–18.

Clifford, James. 1986. "On Ethnographic Allegory." In *Writing Culture: The Poetics and Politics of Ethnography*, edited by James Clifford and George Marcus, 98–121. Berkeley: University of California Press.

Clifford, James. 1988. *The Predicament of Culture*. Cambridge, MA: Harvard University Press.

Clifford, James. 1997. *Routes: Travel and Translation in the Late Twentieth Century*. Cambridge, UK: Harvard University Press.

Clifford, James, and George Marcus, eds. 1986. *Writing Culture: The Poetics and Politics of Ethnography.* Berkeley: University of California Press.

Cohen, Erik. 1971. "Arab Boys and Tourist Girls in a Mixed Jewish-Arab Community." *International Journal of Comparative Sociology* 12: 217–33.

Cohen, Erik. 1972. "Toward a Sociology of International Tourism." *Social Research* 39 (1): 164–82.

Cohen, Erik. 1973. "Nomads from Affluence: Notes on the Phenomenon of Drifter-Tourism." *International Journal of Comparative Sociology* 14 (1): 89–103.

Cohen, Erik. 1974. "Who is a Tourist? A Conceptual Clarification." *Sociological Review* 22 (4): 527–55.

Cohen, Erik. 1979a. "A Phenomenology of Tourist Experiences." *Sociology* 13 (2): 179–201.

Cohen, Erik. 1979b. "Rethinking the Sociology of Tourism." *Annals of Tourism Research* 6 (1): 18–35.

Cohen, Erik. 1988. "Authenticity and Commoditization in Tourism." *Annals of Tourism Research* 15 (3): 371–86.

Cohen, Erik. 1996. *Thai Tourism: Hill Tribes, Islands, and Open-Ended Prostitution.* Bangkok: White Lotus.

Cohen, Erik. 2005. *Contemporary Tourism: Diversity and Change.* Amsterdam: Elsevier.

Cohen, Erik. 2007. "'Authenticity' in Tourism Studies: Aprés la Lutte." *Tourism Recreation Research* 32 (2): 75–82.

Cohen, Erik. 2015. "A Mobilities Approach to Trekking: A 1978 Trek to Hua Mae Kham, Northern Thailand." *International Journal of Tourism Anthropology* 4 (4): 303–26.

Cohen, Erik, and Scott A. Cohen. 2012. "Authentication: Hot and Cool." *Annals of Tourism Research* 39 (3): 1295–314.

Cole, Jennifer. 2001. *Forget Colonialism? Sacrifice and the Art of Memory in Madagascar*. Berkeley: University of California Press.

Cole, Stroma. 2007. "Beyond Authenticity and Commodification." *Annals of Tourism Research* 34 (4): 943–60.

Cole, Stroma. 2008. *Tourism, Culture, and Development: Hopes, Dreams, and Realities*. Clevedon, UK: Channel View.

Collins, Peter, and Anselma Gallinat. 2010. "The Ethnographic Self as Resource: Writing Memory and Experience into Ethnography." In *The Ethnographic Self as Resource*, edited by Peter Collins and Anselma Gallinat, 1–24. Oxford: Berghahn.

Colloredo-Mansfeld, Rudi, Jason Antrosio, and Eric C. Jones. 2011. "Creativity, Place, and Commodities: The Making of Public Economies in Andean Apparel Industries." In *Textile Economies: Power and Value from the Local to the Transnational*, edited by Walter Little and Patricia A. McAnany, 39–55. Walnut Creek, CA: AltaMira.

Comaroff, John L., and Jean Comaroff. 1997. *Of Revelation and Revolution, Volume 2: The Dialectics of Modernity on a South-African Frontier*. Chicago: University of Chicago Press.

Conran, Mary. 2006. "Commentary: Beyond Authenticity: Exploring Intimacy in the Touristic Encounter in Thailand." *Tourism Geographies* 8 (3): 274–85.

Crapanzano, Vincent. 1977. "On the Writing of Ethnography." *Dialectical Anthropology* 2: 69–73.

Crapanzano, Vincent. 1991. "The Postmodern Crisis: Discourse, Parody, Memory." *Cultural Anthropology* 6 (4): 431–46.

Crapanzano, Vincent. 1992. *Hermes' Dilemma and Hamlet's Desire*. Cambridge, MA: Harvard University Press.

Crick, Malcolm. 1985. "Tracing the Anthropological Self: Quizzical Reflections on Field Work, Tourism, and the Ludic." *Social Analysis* 17: 71–92.

Crick, Malcolm. 1989. "Representations of International Tourism in the Social Sciences: Sun, Sex, Sights, Savings, and Servility." *Annual Review of Anthropology* 18: 307–44.

Cripps, Garth. 2009. *Understanding Migration amongst Small-Scale Fishers in Madagascar*. London: Blue Ventures.

Crouch, David. 1999. "Introduction: Encounters in Leisure/Tourism." In *Leisure/Tourism Geographies: Practices and Geographical Knowledge*, edited by David Crouch, 1–16. London: Routledge.

Crouch, David, Lars Aronsson, and Lage Wahlström. 2001. "Tourist Encounters." *Tourist Studies* 1 (3): 253–70.

Crouch, David, and Luke Desforges. 2003. "The Sensuous in the Tourist Encounter." *Tourist Studies* 3 (1): 5–22.

Dahles, Heidi. 1996. "The Social Construction of Mokum: Tourism and the Quest for Local Identity in Amsterdam." In *Coping with Tourists: European Reactions to Mass Tourism,* edited by Jeremy Boissevain, 227–46. Oxford: Berghahn.

Dahles, Heidi, and Karin Bras. 1999. "Entrepreneurs in Romance: Tourism in Indonesia." *Annals of Tourism Research* 26 (2): 267–93.

Dann, Graham S. 1996. "The People of Tourist Brochures." In *The Tourist Image: Myths and Myth Making in Tourism*, edited by Tom Selwyn, 61–81. Chichester, UK: John Wiley.

Davies, James, and Dimitrina Spencer. 2010. "Introduction: Emotions in the Field." In *Emotions in the Field: The Psychology and Anthropology of Fieldwork Experience*, edited by James Davies and Dimitrina Spencer, 1–34. Palo Alto, CA: Stanford University Press.

DeLyser, Dydia. 2001. "When Less is More: Absence and Landscape in a California Ghost Town." In *Textures of Place: Exploring Humanist Geographies*, edited by Paul Adams, Steven Hoelscher, and Karen Till, 24–40. Minneapolis: University of Minnesota Press.

Denzin, Norman. 1974. "The Methodological Implications of Symbolic Interactionism for the Study of Deviance." *The British Journal of Sociology* 25 (3): 269–82.

Denzin, Norman. 1992. *Symbolic Interaction and Cultural Studies.* Oxford: Blackwell.

Derrida, Jacques. 1974. *Of Grammatology*, translated by Gayatri Chakravorty Spivak. Baltimore: Johns Hopkins University Press.

Derrida, Jacques. 2000. *Of Hospitality: Anne Dufourmantelle Invites Jacques Derrida to Respond*, translated by Rachel Bowlby. Palo Alto, CA: Stanford University Press.

Desmond, Jane C. 2001. *Staging Tourism: Bodies on Display from Waikiki to Sea World.* Chicago: University of Chicago Press.

Di Giovine, Michael A. 2009. *The Heritage-scape: UNESCO, World Heritage, and Tourism*. Lanham, MD: Lexington Books.

Di Giovine, Michael A. 2011. "Pilgrimage: Communitas and Contestation, Unity and Difference—An Introduction." *Tourism* 59 (3): 247–69.

Dilthey, Wilhelm. 1976. *Dilthey: Selected Writings,* edited by H. P. Rickman. Cambridge, UK: Cambridge University Press.

Douglas, Mary. 1966. *Purity and Danger*. London: Routledge and Keegan Paul.

Dowler, Lorraine. 2013. "Waging Hospitality: Feminist Geopolitics and Tourism in West Belfast, Northern Ireland." *Geopolitics* 18 (4): 779–99.

Dumont, Jean-Paul. 1991. *Visayan Vignettes: Ethnographic Traces of a Philippine Island*. Chicago: University of Chicago Press.

Dyck, Noel. 2010. "Remembering and the Ethnography of Children's Sports." In *The Ethnographic Self as Resource: Writing Memory and Experience into Ethnography*, edited by Peter Collins and Anselma Gallinat, 150–64. Oxford: Berghahn.

Ebron, Paulla. 1999. "Tourists as Pilgrims: Commercial Fashioning of Transatlantic Politics." *American Ethnologist* 26: 910–32.

Echtner, Charlotte M., and Tazim B. Jamal. 1997. "The Disciplinary Dilemma of Tourism Studies." *Annals of Tourism Research* 24 (4): 868–83.

Edensor, Tim. 1998. *Tourists at the Taj: Performance and Meaning at a Symbolic Site*. London: Routledge.

Edensor, Tim. 2001. "Performing Tourism, Staging Tourism: (Re)producing Tourist Space and Practice." *Tourist Studies* 1 (1): 59–81.

Edwards, Elizabeth. 1996. "Postcards—Greetings from Another World." In *The Tourist Image: Myths and Myth Making in Tourism*, edited by Tom Selwyn, 197–221. Chichester, UK: John Wiley.

Eggan, Fred. 1933. "The Kinship System and Social Organization of the Western Pueblos with Special Reference to the Hopi Indians." PhD diss., University of Chicago.

Eggan, Fred. 1937a. "The Cheyenne and Arapaho Kinship Systems." In *Social Anthropology of North American Tribes,* edited by Fred Eggan, 35–95. Chicago: University of Chicago Press.

Eggan, Fred. 1937b. "Historical Changes in the Choctaw Kinship System." *American Anthropologist* 39 (1): 34–52.

Eggan, Fred. 1954. "Social Anthropology and the Method of Controlled Comparison." *American Anthropologist* 56 (5): 743–63.

Enslin, Elizabeth. 1994. "Beyond Writing: Feminist Practice and the Limitations of Ethnography." *Cultural Anthropology* 9 (4): 537–68.

Eriksen, Thomas H. 1995. *Small Places, Large Issues.* London: Pluto.

Errington, Fred, and Deborah Gewertz. 1989. "Tourism and Anthropology in a Post-Modern World." *Oceania* 27 (3): 37–54.

Evers, Sandra. 2002. *Constructing History, Culture, and Inequality: The Betsileo in the Extreme Southern Highlands of Madagascar*. Leiden: Brill.

Fabian, Johannes. 1983. *Time and the Other: How Anthropology Makes Its Object.* New York: Columbia University Press.

Fauroux, Emmanuel. 2001. "Dynamiques migratoires, tensions Foncières et déforestation dans l'ouest Malgache à la fin du XXème siècle." In *Sociétés paysannes, transitions agraires, et dynamiques écologiques dans le sud-ouest de Madagascar*, edited by Samuel Razanaka, et al., 91–107. Antananarivo: IRD/CNRE Editions.

Featherstone, Mike. 1992. "Postmodernism and the Aestheticization of Everyday Life." In *Modernity and Identity*, edited by Scott Lash and Jonathan Friedman, 265–90. Oxford: Blackwell.

Feifer, Maxine. 1985. *Going Places: The Ways of the Tourist from Imperial Rome to the Present Day*. London: MacMillan.

Feinberg, Benjamin. 2003. *The Devil's Book of Culture: History, Mushrooms, and Caves in Southern Mexico*. Austin: University of Texas Press.

Feinberg, Benjamin. 2006. "'I Was There': Competing Indigenous Imaginaries of the Past and the Future in Oaxaca's Sierra Mazateca." *Journal of Latin American Anthropology* 11 (1): 109–37.

Feldman, Jackie. 1995. "'It Is My Brothers Whom I Am Seeking': Israeli Youths' Pilgrimages to Poland of the Shoah." *Jewish Folklore and Ethnology Review* 17 (1–2): 33–37.

Feldman, Jackie. 2016. *A Jewish Guide in the Holy Land: How Christian Pilgrims Made Me Israeli*. Bloomington: Indiana University Press.

Fernandez, James. 1986. *Persuasions and Performances: The Play of Tropes in Culture.* Bloomington: Indiana University Press.

Fernandez, Nadine T. 1999. “Back to the Future? Women, Race, and Tourism in Cuba.” In *Sun, Sex, and Gold: Tourism and Sex Work in the Caribbean*, edited by Kamala Kempadoo, 81–92. Lanham, MD: Rowman & Littlefield.

Filep, Sebastian, Michael Hughes, Mary Mostafanezhad, and Fiona Wheeler. 2015. “Generation Tourism: Towards a Common Identity.” *Current Issues in Tourism* 18 (6): 511–23.

Fine, Gary. 1993. “The Sad Demise, Mysterious Disappearance, and Glorious Triumph of Symbolic Interactionism.” *Annual Review of Sociology* 19: 61–87.

Firth, Raymond. 1937. *We, the Tikopia: A Sociological Study in Kinship in Primitive Polynesia*. New York: American Book Company.

Forshee, Jill. 1999. “Domains of Pedaling: Souvenirs, *Becak* Drivers, and Tourism in Yogyakarta, Java.” In *Converging Interests: Traders, Travelers, and Tourists in Southeast Asia*, edited by Jill Forshee, Christina Fink, and Sandra Cate, 293–317. Berkeley: International and Area Studies, University of California.

Forshee, Jill. 2001. *Between the Folds: Stories of Cloth, Lives, and Travels from Sumba*. Honolulu: University of Hawai’i Press.

Fosado, Gisela. 2005. “Gay Sex Tourism, Ambiguity, and Transnational Love in Havana.” In *Cuba Transnational*, edited by Damian J. Fernández, 61–78. Gainesville: University Press of Florida.

Foucault, Michel. 1986. “Of Other Spaces.” *Diacritics* 16 (1): 22–27.

Fox, Richard, ed. 1991. *Recapturing Anthropology: Working in the Present*. Santa Fe: SAR Press.

Franklin, Adrian, and Mike Crang. 2001. “The Trouble with Tourism and Travel Theory?” *Tourist Studies* 1 (1): 5–22.

Frey, Nancy. 1998. *Pilgrim Stories: On and Off the Road to Santiago.* Berkeley: University of California Press.

Frohlick, Susan. 2007. “Fluid Exchanges: The Negotiation of Intimacy between Tourist Women and Local Men in a Transnational Town in Caribbean Costa Rica.” *City and Society* 19 (1): 139–68.

Frohlick, Susan. 2010. “The Sex of Tourism? Bodies under Suspicion in Paradise.” In *Thinking through Tourism*, edited by Julie Scott and Tom Selwyn, 51–70. New York: Berg.

Frohlick, Susan. 2013. *Sexuality, Women, and Tourism: Cross-Border Desires through Contemporary Travel*. London: Routledge.

Frohlick, Susan, and Julia Harrison. 2008. “Engaging Ethnography: Researching Tourists.” *Tourist Studies* 8 (1): 5–18.

Fusco, Coco. 1989. “The Border Art Workshop/Taller de Arte Fronterizo: Interview with Guillermo Gómez-Peña and Emily Hicks.” *Third Text* 7: 53–76.

Fusco, Coco. 1997. “Jineteras en Cuba.” *Encuentro de la Cultura Cubana* 4/5: 52–64.

Gable, Eric, and Richard Handler. 1996. “After Authenticity at an American Heritage Site.” *American Anthropologist* 98 (3): 568–78.

Gable, Eric, and Richard Handler. 2005. "Horatio Alger and the Tourist's Quest for Authenticity, or, Optimism, Pessimism, and Middle-Class American Personhood." *Anthropology and Humanism* 30 (2): 124–32.

Garfinkel, Harold. 1967. *Studies in Ethnomethodology*. Englewood, NJ: Prentice-Hall.

Geertz, Clifford. 1966. "Religion as a Cultural System." In *Anthropological Approaches to the Study of Religion*, edited by Michael Banton, 1–46. London: Tavistock. Revised and reprinted as chapter 4 in Geertz 1973.

Geertz, Clifford. 1973. *The Interpretation of Cultures.* New York: Basic Books.

Geertz, Clifford. 1974. "'From the Native's Point of View': On the Nature of Anthropological Understanding." *Bulletin of the American Academy of Arts and Sciences* 28 (1): 26–45. Revised and reprinted in Clifford Geertz, *Local Knowledge: Further Essays in Interpretive Anthropology* (New York: Basic Books, 1983), 55–70.

Geertz, Clifford. 1988. *Works and Lives: The Anthropologist as Author.* Palo Alto, CA: Stanford University Press.

Geertz, Hildred, and Ida Bagus Madé Togog. 2005. *Tales from a Charmed Life: A Balinese Painter Reminisces*. Honolulu: University of Hawai'i Press.

Gibson, Chris. 2010. "Geographies of Tourism: (Un)ethical Encounters." *Progress in Human Geography* 34 (4): 521–27.

Giraudo, Rachel F. 2018. "Performing Our Past to Secure Our Future: A Look at San-Owned Cultural Tourism in the Kalahari." *Senri Ethnological Studies* 99: 215–30.

Gluckman, Max. 1940. "Analysis of a Social Situation in Modern Zululand." *Bantu Studies* 14: 1–30. Reprinted as *Analysis of a Social Situation in Modern Zululand.* Rhodes-Livingstone Papers No. 28. (Manchester, UK: Manchester University Press, 1958).

Goffman, Erving. 1959. *The Presentation of Self in Everyday Life*. New York: Doubleday.

Goffman, Erving. 1961. *Encounters: Two Studies in the Sociology of Interaction.* Indianapolis, IN: Bobbs-Merrill.

Goffman, Erving. 1974. *Frame Analysis: An Essay on the Organization of Experience.* Cambridge, MA: Harvard University Press.

Gómez-Peña, Guillermo. 1993. *Warriors for Gringostroika*. Saint Paul, MN: Graywolf.

Gottlieb, Alma. 1982. "Americans' Vacations." *Annals of Tourism Research* 9 (2): 165–87.

Graburn, Nelson. 1964. "Taqamiut Eskimo Kinship Terminology." Ottawa: Northern Coordination and Research Centre, Greenland Home Rule Government.

Graburn, Nelson. 1967. "The Eskimos and 'Airport Art.'" *Transaction* 4: 28–33.

Graburn, Nelson. 1969. "Art and Acculturative Processes." *International Social Science Journal* 21: 457–68.

Graburn, Nelson, ed. 1976. *Ethnic and Tourist Arts: Cultural Expressions from the Fourth World.* Berkeley: University of California Press.

Graburn, Nelson. 1977. "Tourism: The Sacred Journey." In *Hosts and Guests: The Anthropology of Tourism*, edited by Valene Smith, 17–32. Philadelphia: University of Pennsylvania Press.

Graburn, Nelson. 1980. "Teaching the Anthropology of Tourism." *International Social Science Journal* 32 (1): 56–68.

Graburn, Nelson. 1983a. "The Anthropology of Tourism." *Annals of Tourism Research* 10 (1): 9–33.

Graburn, Nelson. 1983b. *The Anthropology of Tourism.* Special issue of *Annals of Tourism Research* 10 (1).

Graburn, Nelson. 1983c. *To Pray, Pay, and Play: The Cultural Structure of Japanese Domestic Tourism.* Aix-en-Provence: Centre des Hautes Études Touristiques.

Graburn, Nelson. 2002. "The Ethnographic Tourist." In *The Tourist as a Metaphor of the Social World*, edited by Graham S. Dann, 19–39. New York: CABI.

Graburn, Nelson. 2005. "Edward Bruner—Before Batak: Kinship and Change." Paper presented at the annual meeting of the American Anthropological Association, session: "Performance, Tourism, and Ethnographic Practice: An Exploration of the Work of Edward M. Bruner." Washington, DC, November 30–December 4.

Graburn, Nelson. 2008. "Review of *Culture on Tour: Ethnographies of Travel* by Edward M. Bruner." *Annals of Tourism Research* 35 (3): 843–46.

Graeber, David. 2007. *Lost People: Magic and the Legacy of Slavery in Madagascar*. Bloomington: Indiana University Press.

Greenwood, Davydd J. 1972. "Tourism as an Agent of Change: The Spanish Basque Case." *Ethnology* 11 (1): 80–91.

Greenwood, Davydd J. 1977. "Culture by the Pound: An Anthropological Perspective on Tourism as Cultural Commoditization." In *Hosts and Guests: The Anthropology of Tourism*, edited by Valene Smith, 129–37. Philadelphia: University of Pennsylvania Press.

Greenwood, Davydd J. 1982. "Cultural 'Authenticity.'" *Cultural Survival Quarterly* 6 (3): 27–28.

Gregory, Derek. 1994. *Geographical Imaginations*. New York: Blackwell Publishing.

Guldin, Gregory. 1989. "The Anthropological Study Tour in China: A Call for Cultural Guides." *Human Organization* 48 (2): 126–34.

Gupta, Akhil, and James Ferguson. 1997. *Anthropological Locations: Boundaries and Grounds of a Field Science.* Berkeley: University of California Press.

Hales, Rob, Dianne Dredge, Freya Higgins-Desbiolles, and Tazim Jamal. 2018. "Academic Activism in Tourism Studies: Critical Narratives from Four Researchers." *Tourism Analysis* 23 (2): 189–99.

Hallowell, Irving. 1947. "Myth, Culture, and Personality." *American Anthropologist* 49 (4): 544–56.

Handler, Richard. 1988. *Nationalism and the Politics of Culture in Quebec.* Madison: University of Wisconsin Press.

Handler, Richard, and Eric Gable. 1997. *The New History in an Old Museum: Creating the Past at Colonial Williamsburg.* Durham, NC: Duke University Press.

Hannam, Kevin, Mimi Sheller, and John Urry. 2006. "Editorial: Mobilities, Immobilities, and Moorings." *Mobilities* 1 (1): 1–22.

Harkin, Michael. 1995. "Modernist Anthropology and Tourism of the Authentic." *Annals of Tourism Research* 22 (3): 650–70.

Harrison, Julia. 2003. *Being a Tourist: Finding Meaning in Pleasure Travel*. Vancouver: University of British Columbia Press.

Harrison, Julia. 2008. "Shifting Positions." *Tourist Studies* 8 (1): 41–60.

Harrison, Julia. 2012. "'I'm Sorry I Got So Emotional': 'Real' Work and 'Real' Men at the Canadian Cottage." In *Emotion in Motion: Tourism, Affect, and Transformation*, edited by David Picard and Mike Robinson, 231–46. Surrey, UK: Ashgate.

Harrison, Julia. 2013. *A Timeless Place: The Ontario Cottage*. Vancouver: University of British Columbia Press.

Harvey, David. 1990. "Between Space and Time: Reflections on the Geographical Imagination." *Annals of the Association of American Geographers* 80 (3): 418–34.

Hendrickson, Carol. 1995. *Weaving Identities: Construction of Dress and Self in a Highland Guatemala Town*. Austin: University of Texas Press.

Hernández Navarro, Luis. 2008. *The Other Campaign: La otra campaña*. London: Zed.

Herskovits, Melville. 1938. *Acculturation: The Study of Culture Contact*. New York: J. J. Augustin.

Heyns, Ana Cristina Lara. 2014. "Consuming the Zapatista Revolution: A Critical Evaluation of Political Tourism in Southwest Mexico." MA thesis, University of Brighton.

Hicks, Emily. 1991. *Border Writing: The Multidimensional Text*. Minneapolis: University of Minnesota Press.

Himpele, Jeff, and Quetzil E. Castañeda. 1997. *Incidents of Travel in Chichén Itzá*. Ethnographic film, 90 min. Distributed by Documentary Educational Resources, Watertown, MA.

Hitchcock, Michael. 1998. "Tourism, *Taman Mini,* and National Identity." *Indonesia and the Malay World* 26 (75): 124–35.

Hiwasaki, Lisa. 2000. "Ethnic Tourism in Hokkaido and the Shaping of Ainu Identity." *Pacific Affairs* 73 (3): 393–412.

Holland, Dorothy, and Naomi Quinn, eds. 1987. *Cultural Models in Language and Thought.* Cambridge, UK: Cambridge University Press.

Hollinshead, Keith, and Rukeya Suleman. 2017. "Time for Fluid Acumen: A Call for Improved Tourism Studies Dialogue with the Decolonizing World." *Tourism, Culture, & Communication* 17 (1): 61–74.

Huizinga, Johan. 1950. *Homo Ludens: A Study of the Play Element in Culture*. Boston: Beacon.

Hutt, Michael. 1996. "Looking for Shangri-La: From Hilton to Lamichhane." In *The Tourist Image: Myths and Myth Making in Tourism*, edited by Tom Selwyn, 49–60. Chichester, UK: John Wiley.

Ingold, Tim. 2000. *The Perception of the Environment: Essays in Livelihood, Dwelling, and Skill*. London: Routledge.

Jaakson, Reiner. 1986. "Second-Home Domestic Tourism." *Annals of Tourism Research* 13 (3): 367–91.

Jackson, Michael. 2010. "From Anxiety to Method in Anthropological Fieldwork: An Appraisal of George Devereux's Enduring Ideas." In *Emotions in the Field: The*

Psychology and Anthropology of Fieldwork Experience, edited by James Davies and Dimitrina Spencer, 35–54. Palo Alto, CA: Stanford University Press.

Jacobs, Jessica. 2010. *Sex, Tourism, and the Postcolonial Encounter: Landscapes of Longing in Egypt*. London: Ashgate.

Jamal, Tazim, and Mike Robinson, eds. 2009. *The Sage Handbook of Tourism Studies*. London: Sage.

Johnston, Hank. 2000. *Ho! For Yo-Semite.* Yosemite National Park, CA: Yosemite Association.

Jules-Rosette, Benetta. 1984. *The Messages of Tourist Art: An African Semiotic System in Comparative Perspective.* New York: Plenum.

Jules-Rosette, Benetta, and Edward M. Bruner. 1994. "Tourism as Process." *Annals of Tourism Research* 21 (2): 404–6.

Kaell, Hillary. 2014. *Walking Where Jesus Walked: American Christians and Holy Land Pilgrimage.* New York: New York University Press.

Kaeppler, Adrienne L. 1985. "Structured Movement Systems in Tonga." In *Society and the Dance*, edited by Paul Spencer, 92–118. Cambridge, UK: Cambridge University. Press.

Karch, Cecilia, and Graham S. Dann. 1981. "Close Encounters of the Third World." *Human Relations* 34 (4): 249–68.

Kaspin, Deborah. 1997. "Ethnographic Authority and the Tourist Trade: Anthropology in the House of Mirrors." *Anthropological Quarterly* 70 (2): 53–57.

Kaul, Adam. 2004. "At Work in the Field: Problems and Opportunities Associated with Employment during Fieldwork." *Anthropology Matters* 6 (2): 1–9.

Kearney, Michael. 1975. "World View Study and Theory." *Annual Review of Anthropology* 4: 247–70.

Kearney, Michael. 1984. *World View*. Novato, CA: Chandler and Sharp.

Kearney, Michael. 1996. *Reconceptualizing the Peasantry*. London: Routledge.

Keesing, Roger. 1990. "Theories of Culture Revisited." *Canberra Anthropology* 13 (2): 46–90.

Kelly, Walt. 1972. *Pogo: We Have Met the Enemy and He Is Us*. New York: Simon and Schuster.

Kelner, Shaul. 2010. *Tours that Bind: Diaspora, Pilgrimage, and Israeli Birthright Tourism.* New York: New York University Press.

Kenny, Michael, and David Kertzer, eds. 1983. *Urban Life in Mediterranean Europe: Anthropological Perspectives*. Urbana: University of Illinois Press.

Khasnabish, Alex. 2013. *Zapatistas: Rebellion from the Grassroots to the Global.* London: Zed.

Kidder, Alfred. 1930. *Report of the Division of Historical Research,* 91–121. Carnegie Yearbook no. 29 (1929–1930). Washington, DC: Carnegie Institution of Washington.

Kirshenblatt-Gimblett, Barbara. 1998. *Destination Culture: Tourism, Museums, and Heritage*. Berkeley: University of California Press.

Knudsen, Britta Timm, and Anne Marit Waade. 2010a. "Performative Authenticity in Tourism and Spatial Experience: Rethinking the Relations between Travel, Place, and Emotion." In *Re-Investing Authenticity: Tourism, Place, and Emotions,* edited by Britta Timm Knudsen and Anne Marit Waade, 1–19. Bristol: Channel View.

Knudsen, Britta Timm, and Anne Marit Waade, eds. 2010b. *Reinvesting Authenticity: Tourism, Place and Emotions*. Bristol: Channel View.

Kondo, Dorinne. 1986. "Dissolution and Reconstitution of Self: Implications for Anthropological Epistemology." *Cultural Anthropology* 1 (1): 74–88.

Kroeber, Alfred L. 1931. "Historical Construction of Culture Growth and Organic Evolution." *American Anthropologist* 33 (2): 149–56.

Kroeber, Alfred L. 1948. *Anthropology: Culture, Patterns, and Processes*. New York: Harcourt, Brace & World.

Kugelmass, Jack. 1992. "The Rites of the Tribe: American Jewish Tourism in Poland." In *Museums and Communities: The Politics of Public Culture,* edited by Ivan Karp, Christine Mullen Kreamer, and Steven D. Lavine, eds., 382–427. Washington, DC: Smithsonian.

Kummels, Ingrid. 2005. "Love in the Time of Diaspora: Global Markets and Local Meaning in Prostitution, Marriage, and Womanhood in Cuba." *Iberoamericana* 5 (20): 7–26.

Lacy, Julie A., and William A. Douglass. 2002. "Beyond Authenticity: The Meanings and Uses of Cultural Tourism." *Tourist Studies* 2 (1): 5–21.

Lambek, Michael. 2013. "Interminable Disputes in Northwest Madagascar." In *Religion in Disputes: Pervasiveness of Religious Normativity in Disputing Processes*, edited by Franz von Benda-Beckmann et al., 1–18. London: Palgrave Macmillan.

Lassiter, Luke. 2005. *The Chicago Guide to Collaborative Ethnography*. Chicago: University of Chicago Press.

Leddy, Thomas. 1995. "Everyday Surface Aesthetic Qualities: 'Neat,' 'Messy,' 'Clean,' 'Dirty.'" *Journal of Aesthetics and Art Criticism* 53: 259–68.

Lehrer, Erica. 2013. *Jewish Poland Revisited: Heritage Tourism in Unquiet Places.* Bloomington, IN: Indiana University Press.

Leite, Naomi. 2005. "Travels to an Ancestral Past: On Diasporic Tourism, Embodiment and Memory." *Anthropológicas* 9: 273–302.

Leite, Naomi. 2007. "Materializing Absence: Tourists, Surrogates, and the Making of Jewish Portugal." In *Things that Move: The Material Worlds of Tourism and Travel*, edited by Mike Robinson, 1–21. Leeds, UK: Centre for Tourism and Cultural Change.

Leite, Naomi. 2014. "Afterword: Locating Imaginaries in the Anthropology of Tourism." In *Tourism Imaginaries: Anthropological Approaches*, edited by Noel Salazar and Nelson Graburn, 260–78. New York: Berghahn.

Leite, Naomi. 2017. *Unorthodox Kin: Portuguese Marranos and the Global Search for Belonging.* Berkeley: University of California Press.

Leite, Naomi, and Nelson Graburn. 2009. "Anthropological Interventions in Tourism Studies." In *The Sage Handbook of Tourism Studies*, edited by Tazim Jamal and Mike Robinson, 33–64. London: Sage.

Lett, James W., Jr. 1983. "Ludic and Liminoid Aspects of Charter Yacht Tourism in the Caribbean." *Annals of Tourism Research* 10 (1): 35–56.

Lett, James W., Jr. 1989. "Towards a Theory of Tourism: Epilogue." In *Hosts and Guests: The Anthropology of Tourism,* 2nd ed., edited by Valene L. Smith, 275–79. Philadelphia: University of Pennsylvania Press.

Levine, Linda, and David Pizarro. 2004. "Emotion and Memory Research: A Grumpy Overview." *Social Cognition* 22 (5): 530–54.

Lévi-Strauss, Claude. 1970. *Tristes Tropiques*. New York: Atheneum.

Lewis, Charlton T., and Charles Short. 1879. *A Latin Dictionary*. Oxford: Clarendon.

Lill, Elaine. 2011. "Articulation and the Production of Customary Law in Resource Management: A Case Study from Coastal Southwest Madagascar." Paper presented at the Annual Conference of the American Anthropological Association, Montréal, Quebec, Canada, November 21.

Lindquist, Johan. 2008. *The Anxieties of Mobility: Migration and Tourism in the Indonesian Borderlands*. Honolulu: University of Hawai'i Press.

Little, Walter E. 2003. "Performing Tourism: Maya Women's Strategies." *Signs* 29 (2): 527–32.

Little, Walter E. 2004. *Mayas in the Marketplace: Tourism, Globalization, and Cultural Identity*. Austin: University of Texas Press.

Little, Walter E. 2008. "Living within the Mundo Maya Development Project: Strategies of Maya Handicraft Vendors." *Latin American Perspectives* 35 (3): 87–102.

Loeb, Lawrence. 1977. "Creating Antiques for Fun and Profit: Encounters between Iranian Jewish Merchants and Touring Coreligionists." In *Hosts and Guests: The Anthropology of Tourism*, edited by Valene Smith, 185–92. Philadelphia: University of Pennsylvania Press.

Loseke, D. R. 1999. *Thinking about Social Problems: A Constructionist Perspective.* New York: Aldine de Gruyter.

Louie, Andrea. 1996. "Re-Negotiating and Re-Rooting Chinese Identities in the 'Diaspora.'" PhD diss., University of California, Berkeley.

Louie, Andrea. 2004. *Chineseness across Borders: Renegotiating Chinese Identities in China and the United States.* Durham, NC: Duke University Press.

Lowie, Robert. 1935. *The Crow Indians*. Lincoln, NE: University of Nebraska Press.

Lundgren, Silje. 2011. "Heterosexual Havana: Ideals and Hierarchies of Gender and Sexuality in Contemporary Cuba." PhD diss., Uppsala University.

Lynn, Andrea. 2005. "Scholar: Tourists Should Reflect on Themselves, Tourism Experience." *Illinois News Bureau* website (University of Illinois), Research News section, May 11. https://news.illinois.edu/blog/view/6367/207284, accessed July 11, 2016.

Lyotard, Jean-Francois. 1979. *The Postmodern Tradition*, translated by Geoff Bennington and Brian Massumi. Minneapolis: University of Minnesota Press.

MacCannell, Dean. 1973. "Staged Authenticity: Arrangements of Social Space in Tourist Settings." *The American Journal of Sociology* 79 (3): 589–603.

MacCannell, Dean. 1976. *The Tourist: A New Theory of the Leisure Class*. New York: Schocken.

MacCannell, Dean. 1989. *The Tourist: A New Theory of the Leisure Class*, 2nd ed. New York: Schocken.

MacCannell, Dean. 1992. *Empty Meeting Grounds: The Tourist Papers*. London: Routledge.

MacCannell, Dean. 2008. "Why It Never Really Was about Authenticity." *Society* 45 (4): 334–37.

MacCannell, Dean. 2011. *The Ethics of Sightseeing*. Berkeley: University of California Press.

MacClancy, Jeremy. 2002a. "Introduction: Taking People Seriously." In *Exotic No More: Anthropology on the Front Lines,* edited by Jeremy MacClancy, 1–14. Chicago: University of Chicago Press.

MacClancy, Jeremy. 2002b. "Paradise Postponed: The Predicaments of Tourism." In *Exotic No More: Anthropology on the Front Lines,* edited by Jeremy MacClancy, 418–29. Chicago: University of Chicago Press.

Macdonald, Sharon. 1997. "A People's Story: Heritage, Identity, and Authenticity." In *Touring Cultures: Transformations of Travel and Theory*, edited by Chris Rojek and John Urry, 155–75. London: Routledge.

Malinowski, Bronisław. 1944. *A Scientific Theory of Culture and Other Essays*. Chapel Hill: University of North Carolina Press.

Malinowski, Bronisław. 1945. *The Dynamics of Cultural Change: An Inquiry into Race Relations in Africa.* New Haven, CT: Yale University Press.

Marcus, George. 1995. "Ethnography in/of the World System: The Emergence of Multi-Sited Ethnography." *Annual Review of Anthropology* 24: 95–117.

Marcus, George. 2008. "The End(s) of Ethnography: Social/Cultural Anthropology's Signature Form of Producing Knowledge in Transition." *Cultural Anthropology* 23 (1): 1–14.

Marcus, George, and Michael Fischer. 1986. *Anthropology as Cultural Critique: An Experimental Moment in the Human Sciences.* Chicago: University of Chicago Press.

Markula, Pirkko. 1997. "As a Tourist in Tahiti: An Analysis of Personal Experience." *Journal of Contemporary Ethnography* 26 (2): 202–24.

Marschall, Sabine, ed. 2017. *Tourism and Memories of Home: Migrants, Displaced People, Exiles, and Diasporic Communities*. Bristol, UK: Channel View.

Martin, Emily. 1994. *Flexible Bodies: Tracking Immunity from the Days of Polio to the Age of AIDS*. Boston: Beacon.

Maskens, Maïté, and Ruy Blanes. 2016. "Introduction: Ethnography and the Mutualizing Utopia." *Journal of the Anthropological Society of Oxford* 8 (1): 125–42.

Mauss, Marcel. 1985 [1938]. "A Category of the Human Mind: The Notion of Person; The Notion of Self." In *The Category of the Person: Anthropology, Philosophy, History*, edited by Michael Carrithers, Steven Collin, and Steven Lukes, 1–25. Cambridge, UK: Cambridge University Press.

Mauss, Marcel. 2016 [1925]. *The Gift*, edited and translated by Jane I. Guyer. Chicago: HAU Books.

McCloskey, Kathy. 2002. *Swept under the Rug: A Hidden History of Navajo Weaving*. Albuquerque: University of New Mexico Press.

McFarlane, Colin. 2004. "Geographical Imaginations and Spaces of Political Engagement: Examples from the Indian Alliance." *Antipode* 36 (5): 890–916.

McKean, Philip. 1977. "Towards a Theoretical Analysis of Tourism: Economic Dualism and Cultural Involution in Bali." In *Hosts and Guests: The Anthropology of Tourism*, edited by Valene Smith, 93–108. Philadelphia: University of Pennsylvania Press.

Mead, George H. 1934. *Mind, Self, and Society*. Chicago: University of Chicago Press.

Mead, Margaret. 1935. *Sex and Temperament in Three Primitive Societies.* New York: William Morrow.

Medina, Laurie Kroshus. 2003. "Commoditizing Culture: Tourism and Maya Identity." *Annals of Tourism Research* 30: 353–68.

Medley, Steven. 2002. *The Complete Guidebook to Yosemite National Park.* El Portal, CA: The Yosemite Association.

Meisch, Lynn. 1995. "Gringas and Otavaleños: Changing Tourist Relations." *Annals of Tourism Research* 22 (2): 441–62.

Meiu, George Paul. 2017. *Ethno-erotic Economies: Sexuality, Money, and Belonging in Kenya.* Chicago: University of Chicago Press.

Merton, Robert. 1949. *Social Theory and Social Structure: Toward the Codification of Theory and Research.* Glencoe, IL: Free Press.

Michel, Franck. 1998. "Le tourisme international: Une bouée de sauvetage pour Cuba?" In *Tourismes, Touristes, Sociétés*, edited by Franck Michel, 251–87. Paris: L'Harmattan.

Middleton, Karen. 1999. "Introduction." In *Ancestors, Power, and History in Madagascar*, edited by Karen Middleton, 1–36. Leiden: Brill.

Moeran, Brian. 1983. "The Language of Japanese Tourism." *Annals of Tourism Research* 10 (1): 93–108.

Moore, Henrietta L. 2011. *Still Life: Hopes, Desires, and Satisfactions*. Cambridge, UK: Polity.

Morgan, Lewis H. 1877. *Ancient Society; or, Researches in the Lines of Human Progress from Savagery through Barbarism to Civilization*. Chicago: C. H. Kerr.

Morgan, Nigel, Annette Pritchard, Senija Causevic, and Lynn Minnaert. 2018. "Ten Years of Critical Tourism Studies: Reflections on the Road Less Traveled." *Tourism Analysis* 23 (2): 183–87.

Morris-Reich, Amos. 2008. *The Quest for Jewish Assimilation in Modern Social Science*. New York: Routledge.

Mostafanezhad, Mary, and Kevin Hannam. 2014. "Introducing Moral Encounters in Tourism." In *Moral Encounters in Tourism*, edited by Mary Mostafanezhad and Kevin Hannam, 1–12. London: Ashgate.

Mostafanezhad, Mary, and Roger Norum. 2016. "Towards a Geopolitics of Tourism." *Annals of Tourism Research* 61: 226–28.

Mostafanezhad, Mary, Roger Norum, Eric J. Shelton, and Anna Thompson-Carr, eds. 2016. *Political Ecology of Tourism: Community, Power, and the Environment.* New York: Routledge.

Muñoz Ramírez, Gloria. 2008. *The Fire and the Word: A History of the Zapatista Movement and Subcomandante Insurgente Marcos*. London: Zed.

Myerhoff, Barbara. 1978. *Number Our Days*. New York: Simon and Schuster.

Myers, Fred, and Barbara Kirshenblatt-Gimblett. 2001. "Art and Material Culture: A Conversation with Annette Weiner." In *The Empire of Things: Regimes of Value and Material Culture*, edited by Fred Myers, 269–313. Santa Fe: SAR Press.

Nader, Laura. 1972. "Up the Anthropologist—Perspectives Gained from Studying Up." In *Reinventing Anthropology*, edited by Dell Hymes, 284–311. New York: Pantheon.

Naficy, Hamid. 1993. *The Making of Exile Cultures: Iranian Television in Los Angeles.* Minneapolis: University of Minnesota Press.

Nash, Dennison. 1981. "Tourism as an Anthropological Subject." *Current Anthropology* 22 (5): 461–81.

Nash, Dennison. 1996. *The Anthropology of Tourism.* Bingley, UK: Emerald.

Nash, Dennison, ed. 2007. *The Study of Tourism: Anthropological and Sociological Beginnings*. Amsterdam: Elsevier.

Nash, Dennison, and Valene Smith. 1991. "Anthropology and Tourism." *Annals of Tourism Research* 18 (1): 12–25.

Nelson, Rebecca. 2009. "'A Radical-Chic Tourist Attraction': The Impact of Zapatourism on the Project of the Zapatista Army of National Liberation." Paper presented at the 2nd Annual Connecticut Latin Americanist Partnership Conference, West Hartford, CT, December 4.

Ness, Sally A. 1992. *Body, Movement, and Culture: Kinesthetic and Visual Symbolism in a Philippine Community*. Philadelphia: University of Pennsylvania Press.

Ness, Sally A. 1996. "Dancing in the Field: Notes from Memory." In *Corporealities: Dancing Knowledge, Culture, and Power*, edited by Susan Foster, 133–58. Chicago: University of Chicago Press.

Ness, Sally A. 2002. *Where Asia Smiles: An Ethnography of Philippine Tourism.* Philadelphia: University of Pennsylvania Press.

Ness, Sally A. 2011. "Bouldering in Yosemite: Emergent Signs of Place and Landscape." *American Anthropologist* 113 (1): 71–87.

Ness, Sally A. 2015. "A Slip on the Cables: Touristic Rituals and Landscape Performance in Yosemite National Park." In *The Oxford Handbook of Dance and Theatre,* edited by Nadine George-Graves, 452–79. New York: Oxford University Press.

Ness, Sally A. 2016. *Choreographies of Landscape: Signs of Performance in Yosemite National Park.* Oxford: Berghahn.

Norum, Roger, and Mary Mostafanezhad. 2016. "A Chronopolitics of Tourism." *Geoforum* 77: 157–60.

Nuñez, Theron A. 1989. "Touristic Studies in Anthropological Perspective." In *Hosts and Guests: The Anthropology of Tourism,* 2nd ed., edited by Valene Smith, 265–74. University of Pennsylvania Press.

Okley, Judith. 1996. *Own or Other Culture.* London: Routledge.

O'Lear, Shannon. 2018. *Environmental Geopolitics*. London: Rowman & Littlefield.

Olsen, Kjell. 2002. "Authenticity as a Concept in Tourism Research: The Social Organization of the Experience of Authenticity." *Tourist Studies* 2 (2): 159–82.

Ong, Aihwa. 1993. "On the Edge of Empires: Flexible Citizenship among Chinese in Diaspora." *positions* 1 (3): 745–78.

Ortner, Sherry. 1999. "Introduction." In *The Fate of "Culture": Geertz and Beyond*, edited by Sherry Ortner, 1–14. Berkeley: University of California Press.

Overing Kaplan, Joanna. 1981. "Review Article: Amazonian Anthropology." *Journal of Latin American Studies* 13 (1): 151–64.

Pascal, Benjamin. 2008. *De la «Terre des Ancêtres»* aux *Territoires des Vivants: Les enjeux locaux de la gouvernance sur le littoral sud-ouest de Madagascar*. Paris: Museum National d'Histoire Naturelle.

Passariello, Phyllis. 1983. "Never on Sunday?: Mexican Tourists at the Beach." *Annals of Tourism Research* 10 (1): 109–22.

Patico, Jennifer. 2009. "For Love, Money, or Normalcy: Meanings of Strategy and Sentiment in the Russian-American Matchmaking Industry." *Ethnos* 74 (3): 307–30.

Peirce Edition Project, ed. 1998. *The Essential Peirce, Volume 2. Selected Philosophical Writings (1893–1913)*. Bloomington: Indiana University Press.

Pemberton, John. 1994. "Recollections from 'Beautiful Indonesia' (Somewhere beyond the Postmodern)." *Public Culture* 6 (2): 241–62.

Pernecky, Tomas. 2012. "Constructionism: Critical Pointers for Tourism Studies." *Annals of Tourism Research* 39 (2): 1116–37.

Perrotet, Tony. 2002. *Route 66 A.D.: On the Trail of Ancient Roman Tourists*. New York: Random House.

Pfaffenberger, Bryan. 1983. "Serious Pilgrims and Frivolous Tourists: The Chimera of Tourism in the Pilgrimages of Sri Lanka." *Annals of Tourism Research* 10 (1): 57–74.

Picard, David. 2007. "Friction in a Tourism Contact Zone." *Suomen Antropologi* 32 (2): 96–109.

Picard, David. 2011. *Tourism, Magic, and Modernity: Cultivating the Human Garden*. Oxford: Berghahn.

Picard, David. 2012. "Tourism, Awe, and Inner Journeys." In *Emotion in Motion: Tourism, Affect, and Transformation*, edited by David Picard and Mike Robinson, 1–20. Surrey, UK: Ashgate.

Picard, David, and Catarina Nascimento Moreira. 2015. "Hospitality and Dispossession in Madagascar." *Anthropology Today* 31 (6): 12–15.

Picard, David, and Catarina Nascimento Moreira. 2016. "Colonial Memory, Hospitality, and Tourism in Southwestern Madagascar." *International Journal of Tourism Anthropology* 5 (3–4): 187–203.

Picard, David, and Mike Robinson, eds. 2012. *Emotion in Motion: Tourism, Affect, and Transformation*. Surrey, UK: Ashgate.

Picard, David, and Dennis Zuev. 2014. "The Tourist Plot: Antarctica and The Modernity of Nature." *Annals of Tourism Research* 45: 102–15.

Picard, Michel. 1996. *Bali: Cultural Tourism and Tourist Culture.* Singapore: Archipelago.

Picard, Michel. 2008. "Balinese Identity as Tourist Attraction: From 'Cultural Tourism' (*Pariwisata Budaya*) to 'Bali Erect' (*Ajeg Bali*)." *Tourist Studies* 8 (2): 155–73.

Piot, Charles. 2010. *Nostalgia for the Future: West Africa after the Cold War*. Chicago: University of Chicago Press.

Piscitelli, Adriana. 2013. *Transitos: Brasileiras nos mercados transnacionais do sexo*. Rio de Janeiro: Editora da Universidade do Estado do Rio de Janeiro.

Pitarch, Pedro. 2004. "The Zapatistas and the Art of Ventriloquism." *Journal of Human Rights* 3 (3): 291–312.

Pitkänen, K. 2008. "Second-Home Landscape: The Meaning(s) of Landscape for Second-Home Tourism in Finnish Lakeland." *Tourism Geographies* 10 (2): 169–92.

Pitt-Rivers, Julian. 1954. *The People of the Sierra*. London: Weidenfeld and Nicolson.

Pitt-Rivers, Julian. 1977. *The Fate of Shechem, or the Politics of Sex: Essays in the Anthropology of the Mediterranean*. Cambridge, UK: Cambridge University Press.

Pratt, Mary Louise. 1992. *Imperial Eyes: Travel Writing and Transculturation*. Oxford: Routledge.

Pruitt, Deborah, and Suzanne LaFont. 1995. "For Love and Money: Romance Tourism in Jamaica." *Annals of Tourism Research* 22 (2): 422–40.

Rabinow, Paul. 1977. *Reflections on Fieldwork in Morocco*. Berkeley: University of California Press.

Rabinow, Paul, and William Sullivan, eds. 1979. *Interpretive Social Science*. Berkeley: University of California Press.

Radcliffe-Brown, Alfred R. 1952. *Structure and Function in Primitive Society*. London: Cohen and West.

Randrianantenaina, Etienne G. 2009. *Le rite du "Sorognanake Vezo": Espace pour une reflexion théologico-pastorale sur le rite baptismal*. Antananarivo, Madagascar: Institut Catholique de Madagascar.

Rapport, Nigel, and Joanna Overing. 2013. "Alterity." In *Social and Cultural Anthropology: The Key Concepts*, edited by Nigel Rapport and Joanna Overing, 9–17. London: Routledge.

Redfield, Robert. 1941. *The Folk Culture of Yucatan*. Chicago: University of Chicago Press.

Redfield, Robert. 1952. "The Primitive World View." *Proceedings of the American Philosophical Society* 96: 30–36.

Redfield, Robert, Ralph Linton, and Melville Herskovits. 1935a. "Acculturation." *Oceania* 6 (2): 229–33.

Redfield, Robert, Ralph Linton, and Melville J. Herskovits. 1935b. "A Memorandum for the Study of Acculturation." *Man* 35: 145–48.

Redfield, Robert, Ralph Linton, and Melville J. Herskovits. 1936a. "A Memorandum for the Study of Acculturation." *American Anthropologist* 38 (1): 149–52.

Redfield, Robert, Ralph Linton, and Melville Herskovits. 1936b. "The Study of Acculturation." *Africa* 9 (1): 114–18.

Régi, Tamás. 2013. "The Art of the Weak: Tourist Encounters in East Africa." *Tourist Studies* 13 (1): 99–118.

Reisinger, Yvette, and Carol J. Steiner. 2006. "Reconceptualizing Object Authenticity." *Annals of Tourism Research* 33 (1): 65–86.

Robinson, Mike. 2012. "The Emotional Tourist." In *Emotion in Motion: Tourism, Affect, and Transformation*, edited by David Picard and Mike Robinson, 21–46. Surrey, UK: Ashgate.

Roland, Kaifa L. 2011. *Cuban Color in Tourism and La Lucha: An Ethnography of Racial Meaning*. New York: Oxford University Press.

Rosaldo, Renato. 1984. "Grief and a Head-Hunter's Rage: On the Cultural Force of Emotions." In *Text, Play, and Story: The Construction and Reconstruction of Self and Society*, edited by Edward M. Bruner, 178–95. Washington, DC: American Ethnological Society.

Rosaldo, Renato. 1986. "Ilongot Hunting as Story and Experience." In *The Anthropology of Experience*, edited by Victor Turner and Edward M. Bruner, 97–138. Urbana: University of Illinois Press.

Rosaldo, Renato. 1989. *Culture and Truth: The Remaking of Social Analysis*. Boston: Beacon.

Ross, John. 2006. *Zapatistas: Making Another World Possible. Chronicles of Resistance, 2000–2006*. New York: Nation.

Rouse, Roger. 1991. "Mexican Migration and the Social Space of Postmodernism." *Diaspora* 1 (1): 8–23.

Russell, Carl P. 1992. *One Hundred Years in Yosemite*. Yosemite National Park, CA: Yosemite Association.

Said, Edward. 1993. *Culture and Imperialism*. New York: Vintage Books.

Salazar, Noel B. 2010. *Envisioning Eden: Mobilizing Imaginaries in Tourism and Beyond*. Oxford: Berghahn.

Salazar, Noel B. 2012. "Tourism Imaginaries: A Conceptual Approach." *Annals of Tourism Research* 39 (2): 863–82.

Salazar, Noel, and Nelson Graburn, eds. 2014. *Tourism Imaginaries: Anthropological Approaches*. Oxford: Berghahn.

Salomon, Christine. 2009. "Antiquaires et businessmen de la Petite Côte du Sénégal: Le commerce des illusions amoureuses." *Cahiers d'Études Africaines* 193/194 (1–2): 147–76.

Sánchez Taylor, Jacqueline. 2000. "Tourism and 'Embodied' Commodities: Sex Tourism in the Caribbean." In *Tourism and Sex: Culture, Commerce, and Coercion*, edited by Stephen Clift and Simon Carter, 41–53. London: Pinter.

Sanjek, Roger, ed. 1990. *Fieldnotes: The Makings of Anthropology*. Ithaca, NY: Cornell University Press.

Santos, Paula Mota. 2012. "The Power of Knowledge: Tourism and the Production of Heritage in Porto's Old City." *International Journal of Heritage Studies* 18 (5): 444–58.

Santos, Paula Mota. 2017. "Tourism and the Critical Cosmopolitanism Imagination: 'The Worst Tours' in a European World Heritage City." *International Journal of Heritage Studies*. Available online: http://doi.org/10.1080/13527258.2017.1413676

Sapir, Edward. 1916. *Time Perspective in Aboriginal American Culture: A Study in Method*. Ottawa, Canada: Government Printing Bureau.

Sargent, Shirley. 2000. *Yosemite's Innkeepers: The Story of a Great Park and Its Chief Concessionaires*, 2nd ed. Yosemite, CA: Ponderosa.

Sarkissian, Margaret. 2000. *D'Albuquerque's Children: Performing Tradition in Malaysia's Portuguese Settlement*. Chicago: University of Chicago Press.

Saroengallo, Tino. 2010. *Ayah Anak Beda Warna! Anak Toraja Kota Menggugat.* Yogyakarta, Indonesia: Tembi Rumah Budaya.

Schechner, Richard. 1985. *Between Theatre and Anthropology.* Philadelphia: University of Pennsylvania Press.

Schlossberg, Pavel. 2016. *Crafting Identity: Transnational Indian Arts and the Politics of Race in Central Mexico*. Tucson: University of Arizona Press.

Schütz, Alfred. 1962. *Collected Papers I: The Problem of Social Reality*, edited by M. A. Natanson and H. L. van Breda. Dordrecht, The Netherlands: Martinus Nijhoff.

Schwartz, Joan, and James Ryan, eds. 2003. *Picturing Place: Photography and the Geographical Imagination*. New York: I. B. Tauris.

Scott, Julie, and Tom Selwyn. 2010. "Introduction: Thinking through Tourism—Framing the Volume." In *Thinking through Tourism*, edited by Julie Scott and Tom Selwyn, 1–26. New York: Berg.

Selwyn, Tom, ed. 1996. *The Tourist Image: Myths and Myth-Making in Tourism*. Chichester, UK: John Wiley.

Sharp, Lesley A. 1996. *The Possessed and the Dispossessed: Spirits, Identity, and Power in a Madagascar Migrant Town*. Berkeley: University of California Press.

Sheller, Mimi, and John Urry. 2006. "The New Mobilities Paradigm." *Environment and Planning A: Economy and Space* 38 (2): 207–26.

Shepherd, Robert. 2003. "Fieldwork without Remorse: Travel Desires in a Tourist World." *Consumption, Markets, and Culture* 6 (2): 133–44.

Silvey, Rachel. 2000. "Diasporic Subjects: Gender and Mobility in South Sulawesi." *Women's Studies International Forum* 23 (4): 501–15.

Silvey, Rachael. 2006. "Consuming the Transnational Family: Indonesian Migrant Domestic Workers to Saudi Arabia." *Global Networks* 6 (1): 23–40.

Simmel, Georg. 1950. "The Stranger." In *The Sociology of Georg Simmel,* edited and translated by Kurt Wolff, 402–8. New York: Free Press.

Simoni, Valerio. 2013. "Intimate Stereotypes: The Vicissitudes of Being *Caliente* in Touristic Cuba." *Civilisations: Revue Internationale d'Anthropologie et de Sciences Humaines* 62 (1/2): 181–97.

Simoni, Valerio. 2014. "From Tourist to Person: The Value of Intimacy in Touristic Cuba." *Journal of Tourism and Cultural Change* 12 (3): 280–92.

Simoni, Valerio. 2015. "Intimacy and Belonging in Cuban Tourism and Migration." *The Cambridge Journal of Anthropology* 33 (2): 26–41.

Simoni, Valerio. 2016a. "Ethnography, Mutuality, and the Utopia of Love and Friendship in Touristic Cuba." *Journal of the Anthropological Society of Oxford* 8 (1): 143–67.

Simoni, Valerio. 2016b. *Tourism and Informal Encounters in Cuba*. New York: Berghahn.

Simoni, Valerio. Forthcoming. "Encountering Tourism." In *Oxford Research Encyclopedia of Anthropology,* ed. Mark Aldenderfer. Oxford: Oxford University Press.

Simoni, Valerio, and Scott McCabe. 2008. "From Ethnographers to Tourists and Back Again: On Positioning Issues in the Anthropology of Tourism." *Civilisations* 57: 173–89.

Singer, Milton. 1972. *When a Great Tradition Modernizes: An Anthropological Approach to Indian Civilization.* London: Praeger.

Skinner, Jonathan. 2010. "Leading Questions and Body Memories: A Case of Phenomenology and Physical Ethnography in the Dance Interview." In *The Ethnographic Self as Resource*, edited by Peter Collins and Anselma Gallinat, 111–28. Oxford: Berghahn.

Skinner, Jonathan, and Dimitrios Theodossopoulos, eds. 2011. *Great Expectations: Imagination and Anticipation in Tourism*. Oxford: Berghahn.

Smith, M. Estellie. 1982. "The Process of Sociocultural Continuity." *Current Anthropology* 23 (2): 127–42.

Smith, Valene L., ed. 1977. *Hosts and Guests: The Anthropology of Tourism*. Philadelphia: University of Pennsylvania Press.

Smith, Valene L. 2015. *Stereopticon: An Entry to a Life of Travel and Tourism Research*. Putnam Valley, NY: Cognizant Communication Corporation.

Sousa e Santos, Dina de. 2009. "Reading beyond the Love Lines: Examining Cuban *Jineteras*' Discourses of Love for Europeans." *Mobilities* 4 (3): 407–26.

Stanton, Max E. 1977. "The Polynesian Cultural Center: A Multi-Ethnic Model of Seven Pacific Cultures." In *Hosts and Guests: The Anthropology of Tourism*, edited by Valene Smith, 193–206. Philadelphia: University of Pennsylvania Press.

Stasch, Rupert. 2017. "Tourism." In *The Cambridge Encyclopedia of Anthropology*, edited by Feliz Stein, Sian Lazar, Matei Candea, Hildegard Diemberger, Joel Robbins, Andrew Sanchez, and Rupert Stasch. Available online: http://doi.org/10.29164/17tourism.

Steel, Griet. 2012. "Local Encounters with Globetrotters: Tourism's Potential for Street Vendors in Cusco, Peru." *Annals of Tourism Research* 39 (2): 601–19.

Steiner, Christopher B. 1994. *African Art in Transit*. Cambridge, UK: University of Cambridge Press.

Steward, Julian. 1950. *Area Research, Theory and Practice*. New York: Social Science Research Council.

Stjernfelt, Frederik. 2014. *Natural Proposition: The Actuality of Peirce's Doctrine of Dicisigns.* Boston, MA: Docent Press.

Stocking, George W., Jr. 1979. *Anthropology at Chicago: Tradition, Discipline, Department. An Exhibition Marking the Fiftieth Anniversary of the Department of Anthropology, October 1979–February 1980.* Chicago: The Joseph Regenstein Library, University of Chicago.

Stoller, Paul, and Cheryl Olkes. 1987. *In Sorcery's Shadow*. Chicago: University of Chicago Press.

Strauss, Claudia. 2006. "The Imaginary." *Anthropological Theory* 6 (3): 322–44.

Stronza, Amanda. 2001. "Anthropology of Tourism." *Annual Review of Anthropology* 30: 261–83.

Swain, Margaret B. 2016. "Embodying Cosmopolitan Paradigms in Tourism Research." In *Tourism Research Paradigms: Critical and Emergent Knowledges*, edited by Ana Maria Munar and Tazim Jamal, 87–112. Bingley, U.K.: Emerald.

Swain, Margaret B., and Derek Hall. 2007. "Gender Analysis in Tourism: Personal and Global Dialectics." In *The Critical Turn in Tourism Studies: Innovative Research Methodologies*, edited by Irena Ateljevic, Annette Pritchard, and Nigel Morgan, 91–104. Oxford: Elsevier.

Swedo, Suzanne. 2011. *Hiking Yosemite National Park: A Guide to 59 of the Park's Greatest Adventures.* Guilford, CT: Falcon Guides.

Sylvain, Renée. 2005. "Disorderly Development: Globalization and the Idea of 'Culture' in the Kalahari." *American Ethnologist* 32 (3): 354–70.

Tax, Sol. 1935. "Primitive Social Organization with Some Description of the Fox Indians." PhD diss., University of Chicago.

Tax, Sol. 1952. "Action Anthropology." *America Indigena* 12: 103–9.

Tax, Sol. 1953. *Penny Capitalism: A Guatemalan Indian Economy.* Smithsonian Institution, Institute of Social Anthropology, Publication no. 16. Washington, DC: US Government Printing Office.

Theodossopoulos, Dimitrios. 2013. "Introduction: Laying Claim to Authenticity: Five Anthropological Dilemmas." *Anthropological Quarterly* 86 (2): 337–60.

Thrift, Nigel. 1996. *Spatial Formations*. London: Sage.

Thrift, Nigel. 2000a. "Afterwords." *Environment and Planning D: Society and Space* 18 (2): 213–55.

Thrift, Nigel. 2000b. "Still Life in Nearly Present Time: The Object of Nature." *Body & Society* 6 (3/4): 34–57.

Throop, Jason C. 2014. "Friendship as Moral Experience: Ethnographic Dimensions and Ethical Reflections." *Suomen Antropologi* 39 (1): 68–80.

Thurot, Jean Maurice, and Gaétane Thurot. 1983. "The Ideology of Class and Tourism: Confronting the Discourse of Advertising." *Annals of Tourism Research* 10 (1): 173–89.

Timothy, Dallen J. 2018. "Geography: The Substance of Tourism." *Tourism Geographies* 20 (1): 166–69.

Timothy, Dallen J., and Stephen W. Boyd. 2006. "Heritage Tourism in the 21st Century: Valued Traditions and New Perspectives." *Journal of Heritage Tourism* 1 (1): 1–16.

Tribe, John. 1997. "The Indiscipline of Tourism." *Annals of Tourism Research* 24 (3): 638–57.

Tribe, John. 2010. "Tribes, Territories, and Networks in the Tourism Academy." *Annals of Tourism Research* 37 (1): 7–33.

Tribe, John. 2016. "The Truth about Tourism." *Annals of Tourism Research* 33 (2): 360–81.

Tsing, Anna. 1993. *In the Realm of the Diamond Queen.* Princeton, NJ: Princeton University Press.

Tucker, Hazel. 1997. "The Ideal Village: Interactions through Tourism in Central Anatolia." In *Tourists and Tourism: Identifying with People and Places*, edited by Simone Abram, Don MacLeod, and Jackie Waldren, 107–28. Oxford: Berg.

Tucker, Hazel. 1999. "Living with Tourism: Tourism, Identity and Change in a Village in Central Turkey." PhD diss., Durham University.

Tucker, Hazel. 2001. "Tourists and Troglodytes: Negotiating for Sustainability." *Annals of Tourism Research* 28 (4): 868–91.

Tucker, Hazel. 2011. "Success and Access to Knowledge in the Tourist-Local Encounter: Confrontations with the Unexpected in a Turkish Community." In *Great Expectations: Imagination and Anticipation in Tourism,* edited by Jonathan Skinner and Dimitrios Theodossopoulos, 27–39. Oxford: Berghahn.

Turner, Louis, and John Ash. 1975. *The Golden Hordes: International Tourism and the Pleasure Periphery*. New York: St. Martin's Press.

Turner, Victor. 1969. *The Ritual Process: Structure and Anti-Structure.* Chicago: Aldine.

Turner, Victor. 1973. "The Center Out There: Pilgrim's Goal." *History of Religions* 12 (3): 191–230.

Turner, Victor. 1974. *Dramas, Fields, and Metaphors: Symbolic Action in Human Society.* Ithaca, NY: Cornell University Press.

Turner, Victor. 1986. "The Anthropology of Performance." In *The Anthropology of Performance*, by Victor Turner, 72–98. New York: PAJ Publications.

Turner, Victor, and Edward M. Bruner, eds. 1986. *The Anthropology of Experience*. Urbana: University of Illinois Press.

Urry, John. 1990. *The Tourist Gaze: Leisure and Travel in Contemporary Societies*. London: Sage.

Urry, John, and Jonas Larsen. 2011. *The Tourist Gaze 3.0*. London: Sage.

Vail, Margaret. 2004. "Right of Passage: Backpacker Subculture and the 'Gentrification' of Tourism in Bolivia." PhD diss., New York University.

Van den Abbeele, Georges. 1980. "Sightseers: The Tourist as Theorist." *Diacritics* 10: 3–14.

Van den Berghe, Pierre. 1994. *The Quest for the Other: Ethnic Tourism in San Cristobal, Chiapas*. Seattle: University of Washington Press.

Van den Berghe, Pierre, and Charles F. Keyes, eds. 1984. *Tourism and Re-Created Ethnicity*. Special issue of *Annals of Tourism Research* 11 (3).

Van Gennep, Arnold. 1909. *Les Rites de Passage.* Paris: Emile Nourry.

Van Maanen, John. 1988. *Tales of the Field: On Writing Ethnography*. Chicago: University of Chicago Press.

Veijola, Soile, and Eeva Jokinen. 1994. "The Body in Tourism." *Theory, Culture and Society* 11 (3): 125–51.

Venkatesan, Soumhya, Jeanette Edwards, Rane Willerslev, Elizabeth Povinelli, and Perveez Mody. 2011. "The Anthropological Fixation with Reciprocity Leaves No Room for Love: 2009 Meeting of the Group for Debates in Anthropological Theory." *Critique of Anthropology* 31 (3): 210–50.

Vergara-Camus, Leandro. 2014. *Land and Freedom: The MST, the Zapatistas, and Peasant Alternatives to Neoliberalism*. London: Zed.

Visweswaran, Kamala. 1994. *Fictions of Feminist Ethnography.* Minneapolis: University of Minnesota Press.

Volkman, Toby A. 1985. *Feasts of Honor: Ritual and Change in the Toraja Highlands.* Champaign: University of Illinois Press.

Wagner, Ulla. 1977. "Out of Time and Place: Mass Tourism and Charter Trips." *Ethnos* 42 (1–2): 38–52.

Wallace, Tim, ed. 2005. *Tourism and Applied Anthropologists: Linking Theory and Practice.* Hoboken, NJ: Wiley-Blackwell.

Wang, Ning. 1999. "Rethinking Authenticity in Tourism Experience." *Annals of Tourism Research* 26 (2): 349–70.

Waterson, Roxana. 2003. "The Immortality of the House in Tana Toraja." In *The House in Southeast Asia: A Changing Social, Economic, and Political Domain*, edited by Stephen Sparks and Signe Howell, 34–52. London: Routledge Curzon.

Waterson, Roxana. 2009. *Paths and Rivers: Sa'dan Toraja Society in Transformation.* Leiden: KITLV.

Wells, Stacey. 1998. *Yosemite and Sequoia/Kings Canyon National Parks.* New York: MacMillan.

Whitmarish, Ian, and Alexander S. Dent, eds. 2017. *Reciprocal Radicalizing: Commemorating the Work of James A. Boon.* Special issue of *Anthropological Quarterly* 90 (4).

Winter, Tim. 2009. "Conclusion: Recasting Tourism Theory towards an Asian Future." In *Asia on Tour: Exploring the Rise of Asian Tourism*, edited by Tim Winter, Peggy Teo, and T. C. Chang, 315–25. London: Routledge.

Winter, Tim, Peggy Teo, and T. C. Chang, eds. 2009. *Asia on Tour: Exploring the Rise of Asian Tourism.* London: Routledge.

Wolf, Marjory. 1992. *A Thrice-Told Tale: Feminism, Postmodernism, and Ethnographic Responsibility.* Palo Alto, CA: Stanford University Press.

Wood, Robert E. 1998. "Touristic Ethnicity: A Brief Itinerary." *Ethnic and Racial Studies* 21 (2): 218–41.

Wood, William W. 2008. *Made in Mexico: Zapotec Weavers and the Global Ethnic Art Market*. Bloomington: Indiana University Press.

Yamashita, Shinji. 2003. *Bali and Beyond: Explorations in the Anthropology of Tourism.* New York: Berghahn.

Zhu, Yujie. 2012. "Performing Heritage: Rethinking Authenticity in Tourism." *Annals of Tourism Research* 39 (3): 1495–513.

Zorn, Elayne. 2004. *Weaving a Future: Tourism, Cloth, and Culture on an Andean Island*. Iowa City: University of Iowa Press.

Index

Page references for figures are italicized.

About the Contributors

Kathleen M. Adams is Professor of Anthropology at Loyola University Chicago and Adjunct Curator at the Field Museum of Natural History. Her research centers on the critical analysis of tourism, particularly its intersections with identity politics, arts, museums, and heritage. Her books include the award-winning *Art as Politics: Re-crafting Identities, Tourism, and Power in Tana Toraja, Indonesia* (2006), *Indonesia: History, Culture, Heritage* (2019), *Everyday Life in Southeast Asia* (2012, with Kathleen Gillogly), and *Home and Hegemony: Domestic Work and Identity Politics in South and Southeast Asia* (2000, with Sara Dickey). She has also authored articles on various tourism-oriented topics, including domestic tourism and nation-building, public interest anthropology in heritage sites, ethical quandaries in tourism research, and terrorism and tourism.

Edward M. Bruner is Professor Emeritus of Anthropology and of Criticism and Interpretive Theory at the University of Illinois. His publications include *Text, Play, and Story: The Construction and Reconstruction of Self and Society* (1984), *The Anthropology of Experience* (1986, with Victor Turner), and *Culture on Tour: Ethnographies of Travel* (2005). He is past president of the American Ethnological Society and the Society for Humanistic Anthropology.

Margaret Byrne Swain is a cultural anthropologist, retired from Women and Gender Studies at the University of California, Davis. She continues to be active in both Southwest China ethnography and transdisciplinary tourism studies, with particular interest in feminist and critical cosmopolitan theories. From 2010 to 2014 she served as president of the Tourism Research Committee of the International Sociological Association. Besides authoring numerous research articles and book chapters, she conceived of and edited the special issue of *Annals of Tourism Research* on "Gender in Tourism" (1995), and is coeditor of *Gender, Tourism, Fun(?)* (2002) and *Explorers and Scientists in China's Borderlands* (2011).

Quetzil E. Castañeda is Founding Director of OSEA—The Open School of Ethnography and Anthropology, member of the Graduate Faculty at Indiana University, and editor of the *Journal of Latin American and Caribbean Anthropology* (2018–2022). He has published extensively on tourism in relation to archaeology, museums, heritage politics, Maya culture, New Age spiritualism, ethics, and art and handicrafts, among other topics. He is filmmaker (with Jeffrey Himpele) of the award-winning *Incidents of Travel in Chichén Itzá* (DER 1997), a visual ethnography of spiritual and cultural tourism at a Maya archaeological site. He was the founding convener of the American Anthropological Association's Anthropology of Tourism Interest Group in 2013. In 2004 he founded OSEA, an ethnographic field school and field-study-abroad program in Pisté-Chichén that focuses on engaged anthropology via educational tourism and community-based transcultural exchange.

Michael A. Di Giovine is Associate Professor of Anthropology at West Chester University of Pennsylvania and director of its Museum of Anthropology and Archaeology. A former tour operator, he is a member of ICOMOS's International Cultural Tourism Committee (ICTC) and a regional representative for the IUAES Commission on the Anthropology of Tourism. The convenor of the American Anthropological Association's Anthropology of Tourism Interest Group (2016–2019), he has also served on the AAA's Task Force on Cultural Heritage. He is the author of *The Heritage-Scape: UNESCO, World Heritage, and Tourism* (2008) and coeditor of *Edible Identities: Food as Cultural Heritage* (2014, with Ronda Brulotte), *Tourism and the Power of Otherness* (2014, with David Picard), and *The Seductions of Pilgrimage* (2015, with David Picard). The book reviews editor for *Journeys* and the *Journal of Tourism and Cultural Change*, he edits the Lexington Books series *The Anthropology of Tourism: Heritage, Mobility, and Society*.

Nelson Graburn is Professor Emeritus of Anthropology at the University of California–Berkeley. Among his books and special journal issues are *Ethnic and Tourist Arts* (1976), *The Anthropology of Tourism* (1983), *To Pray, Pay and Play: The Cultural Structure of Japanese Domestic Tourism* (1983), *Tourism Social Sciences* (1991), *Tourism and Glocalization: Perspectives in East Asian Studies* (2010), *Tourism Imaginaries: Anthropological Approaches* (2014, with Noel Salazar), *Tourism Imaginaries at the Disciplinary*

Crossroads: Place, Practice, Media (2016, with Maria Gravari Barbas), and *Indigenous Tourism Movements* (2018, with Alexis Bunten). He is a founding member of the International Academy for the Study of Tourism, the Research Committee on Tourism (RC-50) of the International Sociological Association, and the Berkeley Tourism Studies Working Group, and he serves on the editorial board of *Annals of Tourism Research.*

Julia Harrison, DPhil (Oxon) is Professor Emerita in Anthropology at Trent University, Peterborough, Ontario, Canada. Her research areas include Indigenous representation in museums; the nature of tourist interaction and experience; and Ontario second-home domestic tourism. Key publications include *Being a Tourist* (2003), the coedited volume *Historicizing Canadian Anthropology* (2006), and *A Timeless Place: The Ontario Cottage* (2013), in addition to numerous scholarly articles in museum studies, tourism studies, and anthropology journals. Her current research focuses on the gendered nature of the Canadian cottage experience and the experience of postwar British immigrants to Western Canada.

Naomi M. Leite is Assistant Professor in Social Anthropology at SOAS, University of London. Her research takes tourism encounters as a context for exploring questions of identification, imagination, relatedness, belonging, and exclusion across domains and scales of sociality. She is author of the multiple-award-winning ethnography *Unorthodox Kin: Portuguese Marranos and the Global Search for Belonging* (2017) and coeditor of *Home Imaginings: Travel, Place, and Belonging* (2018, with Paula Mota Santos), a special issue of the *International Journal of Tourism Anthropology*. A cofounder of the Berkeley Tourism Studies Working Group and founding co-convener of the American Anthropological Association's Anthropology of Tourism Interest Group (2013–2016), she serves on the Royal Anthropological Institute's Tourism Committee and on the editorial boards of *Annals of Tourism Research* and the *Journal of Tourism and Cultural Change*.

Walter E. Little, Professor of Anthropology at the University at Albany, SUNY, conducts ethnographic research in Guatemala and Mexico with attention to identity politics and political economy, through the exploration of Kaqchikel and K'ichee' Mayas' livelihoods as artisans and vendors in urban heritage sites. His monograph *Mayas in the Marketplace: Tourism, Globalization, and Cultural Identity* (2004) won Best Book of 2005 from the New England Council for Latin American Studies; his coedited volume *Street Economies in the Urban Global South* (2013) won the Society for the Anthropology of Work Book Prize in 2014.

Mary Mostafanezhad is Associate Professor in the Department of Geography and Environment at the University of Hawai'i at Manoa. She is author of *Volunteer Tourism: Popular Humanitarianism in Neoliberal Times* (2014) and coeditor of six books, among them *At Home and in the Field: Ethnographic Encounters in Asia and the Pacific Islands* (2015), *The Political Ecology of Tourism: Community, Power, and the Environment* (2016), and *Geopolitics of Tourism: Assemblages of Power, Mobility, and the State* (2019). She is also cofounder of the Critical Tourism Studies Asia-Pacific network and the American Anthropological Association's Anthropology of Tourism Interest Group, and she serves on numerous editorial boards such as *Tourism Geographies*, *Geopolitics*, and *Geoforum*.

Sally Ann Ness is Professor of Anthropology, University of California, Riverside. She has worked in urban provincial centers in Southeast Asia and in Yosemite National Park, California. She has written on the semiotics of festival life, dance, and sport, as well as on tourism development and its consequences for cultural practice and cultural identity. Her books include *Choreographies of Landscape: Signs of Performance in Yosemite National Park* (2016), *Where Asia Smiles: An Ethnography of Philippine Tourism* (2002), and the award-winning *Body, Movement, and Culture: Kinesthetic and Visual Symbolism in a Philippine Community* (1992). She is coeditor, with Patrick Alcedo and Hendrik Maier, of *Religious Festivals in Contemporary Southeast Asia* (2016) and, with Carrie Noland, of *Migrations of Gesture* (2008).

David Picard is former Professor of Tourism Anthropology at the University of Lausanne, Switzerland, and now a research associate at the Instituto Superior de Agronomia at the University of Lisbon, Portugal. He holds a PhD in anthropology from the University of La Réunion, Indian Ocean. He is coeditor of several books in the anthropology of tourism, among them *Emotion in Motion: Tourism, Affect, and Transformation* (2012, with Mike Robinson); *Couchsurfing Cosmopolitanisms* (2013, with Sonja Buchberger), *Tourism and the Power of Otherness: Seductions of Difference* (2014, with Michael Di Giovine), and author of *Tourism, Magic and Modernity: Cultivating the Human Garden* (2011). His current research is on wine tourism and winemaking.

Valerio Simoni is Senior Research Fellow at the Global Migration Centre and the Department of Anthropology and Sociology, the Graduate Institute, Geneva, Switzerland, and Research Associate at the Instituto Universitário de Lisboa (ISCTE-IUL), Centro em Rede de Investigação em Antropologia, Lisbon, Portugal. He is author of the award-winning *Tourism and Informal*

Encounters in Cuba (2016), as well as numerous contributions to edited volumes and journals in both anthropology and tourism studies including *Anthropological Theory*, *Cambridge Journal of Anthropology*, *Journal of Anthropological Theory*, and *Journal of Tourism and Cultural Change*. With ethnographic field research in Cuba and Spain, his work contributes to scholarship in the social sciences on intimacy, economic practice, morality, tourism, migration, transnationalism, and globalization.

CPSIA information can be obtained
at www.ICGtesting.com
Printed in the USA
LVHW081459230222
711834LV00004B/177

9 781498 516334